I0469430

United States Fire Administration/National Fire Data Center

Fire and the Older Adult

January 2006

 **Homeland Security**

Department of Homeland Security
United States Fire Administration
National Fire Data Center

UNITED STATES FIRE ADMINISTRATION MISSION STATEMENT

As an entity of the Department of Homeland Security, the mission of the USFA is to reduce life and economic losses due to fire and related emergencies, through leadership, advocacy, coordination, and support. We serve the Nation independently, in coordination with other Federal agencies, and in partnership with fire protection and emergency service communities. With a commitment to excellence, we provide public education, training, technology and data initiatives.

CONTENTS

LIST OF FIGURES

LIST OF TABLES

ACKNOWLEDGEMENTS

Fire and the Older Adult is an updated report based on the U.S. Fire Administration's (USFA's) *Fire Risks for the Older Adult*. The cornerstone of this report is the original analysis performed from the National Fire Incident Reporting System (NFIRS). USFA renews its continued gratitude to the fire departments and State fire marshals' offices that provide statistical information to make NFIRS an effective reality.

The report relies heavily on compilations of research performed by other U.S. agencies. The USFA acknowledges the U.S. Census Bureau, the Federal Interagency Forum on Aging Related Statistics, the National Center for Health Statistics, and the Department of Health and Human Services' Administration on Aging for the research, analysis, and general statistics that made this report possible.

The public education section of this report was completed with the help of Sharon Gamache at the National Fire Protection Association, officials at the American Health Care Association, and Janet Lehman at the Florida Injury Prevention Program for Seniors. USFA thanks them for providing campaign materials for evaluation purposes.

The *Fire and the Older Adult* report was prepared by TriData, a division of System Planning Corporation in Arlington, Virginia. The project leader for the report was Patricia Frazier. She was assisted by Jeff Burkeen, Andrew Campanella, Lucius Lamar, Susan Martin, and Elena Moreno.

EXECUTIVE SUMMARY

Fire is a frightening possibility for older adults (65+) and a reality for far too many. Older adults con front distinct fire risk factors every day—many of which do not affect the young. This report delineates those risk factors and presents the statistics regarding the fire problem among the elderly in the United States.

Demographics of Older Adults

In 2000, individuals 65 years and older comprised 12 percent of the America's population. By 2020, the U.S. Census Bureau projects that the proportion of older Americans will rise to 16 percent (55 mil lion older adults) and that by 2050 there will be more than 86 million older Americans, accounting for 21 percent of the U.S. population.

More than half of older Americans are between ages 65 and 74; 88 percent are between 65 and 84. Fifty nine percent of the elderly population are women and 83 percent are white. Seventy three per cent of elderly men are married compared with 41 percent of elderly women.

Geographically, the largest population groups of older Americans reside in California, Florida, New York, Texas, and Pennsylvania. The South and Midwest have the largest number of elderly residents as a percentage of the overall population.

Risk Factors for Older Adults

Older adults are more likely than their younger counterparts to suffer from reduced sensory abilities such as smell, touch, vision, and hearing, and from diminished mental faculties such as dementia, Alzheimer's disease, and depression. Such impairments tend to reduce older adults' reaction times and place them at a higher risk for causing fires, and thus at a higher risk of fire death and fire injury.

Disabilities present additional fire risks and concerns for the elderly. Twenty percent of Medicare en rollees aged 65 and older are unable to complete at least some of the normal activities of daily living (ADL) necessary for a degree of self sufficiency, including bathing, dressing, getting in and out of bed, getting around inside, toileting, and eating.

Economic and social concerns also contribute to the fire risk for older adults. Most live on fixed incomes and at least 10 percent live in poverty. Thus, they may be unable to afford to make necessary home improvements that could substantially reduce their risk of fire. In addition, studies have shown a relationship between income and health; lower income older adults may be at a higher risk of fire because their health is also poor.

Risk Factors for Fire in Long-Term Care Facilities

Long term care facilities provide many opportunities and amenities for their residents and are often a safe, comfortable, and viable living alternative for older Americans. Although fire and building inspections, codes, and trained staff mitigate the fire risk in these facilities somewhat, unique fire risks do exist. These fire risks are largely due to the heightened level of resident impairment, the facility's layout, and the types of materials and equipment present.

The predominant fire risk in long term care facilities is the impaired health of residents. At least 75 percent of nursing home patients need assistance with three or more activities of daily living. The most common medical conditions among residents are mental disorders and diseases of the nervous system and senses. Sixty three percent of long term care patients use wheelchairs.

In assisted living facilities, residents generally require a lesser degree of assistance. Still, 81 percent of residents need help performing at least one ADL, and 52 percent have cognitive impairments.

Patient care devices, cooking equipment, electrical products, and other systems present in long term or assisted living facilities pose fire risks as well. Compressed gases and flammable liquids used for resident care or facility maintenance increase the risk and severity of fire, and are hazardous both to residents and to the fire service during fire situations.

The architectural layout of long term care facilities—combined with the heightened use of wheel chairs and resident mobility impairments—add to the difficulty of fire evacuation. Sprinklers are required in long term care facilities; they are not required in older nursing homes that have been constructed of noncombustible materials. Further, there are no federal standards requiring smoke alarms in individual nursing home rooms.

Risk Factors for Fire in Home Health and Hospice Care

Home health care and hospice care situations are other viable alternatives for older Americans with impairments and health problems. But they too pose distinct fire risks.

Health is again the key risk factor in home health and hospice situations. The most common illnesses among home health care residents are circulatory system diseases, heart disease, and injuries/poison ing. Cancer is the primary disease among residents receiving hospice care. Seventy four percent of old er adults in home care situations require assistance with activities of daily living and instrumental ac tivities of daily living—light housework, meal preparation, getting around outside, managing money, and using the telephone.

The presence of gases, flammable liquids, and electrical devices are even riskier in home health care situations than in long term or assisted living facilities as no enforceable regulations on storage and maintenance of such materials exist. Smoking, often banned or closely monitored in long term care facilities, is harder to control in home health care and is a significant risk factor. As discussed above, home improvements that reduce fire risk may be postponed either because of affordability or because of lack of knowledge or attention.

Risk of Fire Fatalities

Older adults are 2.5 times more likely to die in fires than the overall population. As Americans age, their fire risk increases. There is a relationship between risk, gender, and race, as well. Older men are at a substantially higher risk of fire death than women, and African Americans are at much greater risk of dying in fires than whites.

Residential Fires and Older Adults

Data regarding residential structure fires and older adults show that, despite the differences in fire risk factors, there are many similarities between fires involving the elderly and fires involving the non elderly. But there are also important differences, such as the time of day fatal and injurious fires occur, and the gender and racial breakdown of fire fatalities and injuries.

According to data from the National Fire Incident Reporting System (NFIRS), 34 percent of the people who died in residential structure fires and 14 percent of the people who were injured in 2002 were aged 65+. More elderly men died in residential structure fires in 2002 than did women, but more women were injured. Because females have longer life expectancies, female deaths and injuries increase as the older population ages.

Older adults are more likely to die or be injured in fires during the midmorning and early afternoon than those 18 to 64 years of age, most likely because the elderly are at home during those hours and not working. Deaths and injuries by month differ little between older adults and the 18 to 64 population, with more fatalities and injuries occurring during the winter, and the fewest in the summer and early fall.

The highest percentage of older adults were located in a bedroom at the time of fire death or injury, and the highest percentages of elderly people died or were injured while sleeping, escaping a fire, or attempting to control a fire. For injuries, more older adults were injured escaping fires than attempting to control them, while those aged 18 to 64 were more likely to be injured controlling a fire than escaping it.

The predominant causes of fires in which an older adult was killed are smoking, open flame, heating, and suspicious acts. Cooking, open flames, smoking, and heating caused more fires that resulted in injuries among the elderly than other fire causes.

Fire Education and Prevention

Many organizations address the risk of fires, deaths, and injuries to older adults through information about reducing their fire risk. In addition to the U.S. Fire Administration's public information campaign, *A Fire Safety Campaign for People 50 Plus,* organizations like the Florida Department of Elder Affairs, the American Health Care Association, the National Center for Assisted Living, the National Fire Protection Association, and the American Burn Association have active fire prevention and education programs for the elderly.

In the summer of 2004, the U.S. Fire Administration (USFA) launched its most comprehensive and intensive public fire education campaign for elderly Americans. A FIRE SAFETY CAMPAIGN FOR PEOPLE 50-PLUS provides detailed fire prevention information to assist fire departments and other USFA partners in mitigating the risk of fire fatalities and injuries among the 50 and over population.

This report, *Fire and the Older Adult*, analyzes the fire risk to persons aged 65 and older as a complement to that campaign. The report provides an extensive review of the fire situation for older adults in the United States and evaluates fire risk factors and risks of fire injury and fatality among that population group.

On average, more than 1,000 Americans aged 65 years and older die each year in home fires and more than 2,000 are injured. In 2001 alone, 1,250 older adults died as the result of fire incidents. Moreover, the elderly are 2.5 times more likely to die in a residential fire than the rest of the population. With the U.S. Census Bureau predicting that increases in the senior population will continue to outpace increases in the overall population, the elderly fire problem will undoubtedly grow in importance.

After offering an overview of the U.S. demographics of the 65 and older population, this report discusses how physical, emotional, social, economic, and residential factors have unique impacts on seniors. Fire data from the National Center for Health Statistics (NCHS) and the National Fire Incident Reporting System (NFIRS) are then analyzed to assess the nature of the fire problem for older adults. Information on several regional, national, and private/public information programs designed to lower fire injuries and fatalities among Americans 65 and over is also provided in the final section.

This report relies on many sources, including NFIRS, NCHS, and the U.S. Census Bureau as well as information from a variety of public and private organizations.

The USFA is committed to lowering the number of fire fatalities and injuries among people 65 years and older. *Fire and the Older Adult* provides concerned citizens, organizations, and USFA partners with the research needed to effectively combat the U.S. fire problem for this high risk population segment.

This report relies on data from many sources and organizations. This chapter briefly describes the major data sources used and presents the methodology that was applied in handling the "unknowns" and "undetermined" entries in the NFIRS database.

Residential structure fires are the focal point of discussion as they account for the majority of both numbers of fires and casualties. In this report, older adults are defined as individuals aged 65 and older. In some of the analyses, members of this age group are divided into subgroups to show variations in trends and patterns as people age. For purposes of this report, the terms *older adult*, *senior*, *65+*, and *elderly* all refer to individuals aged 65 or older. The term *oldest old* generally refers to the oldest segment of the population, those aged 85 or older.

Major Data Sources

National Fire Incident Reporting System. The fire related findings in this report are based primarily on analysis of NFIRS fire incident data for 2002. NFIRS is a voluntary data collection system adminis tered by the USFA, a division under the Department of Homeland Security (DHS). The participating fire departments include career, volunteer, and combination departments that serve communities ranging from rural hamlets to the largest cities. Participation in NFIRS is State based and voluntary. Not all States participate and, for those that do, reported fire incidents do not reflect the entirety of a State's fire activity. Also, not all recorded information is complete. Nevertheless, each year of NFIRS data contains between 540,000 and 800,000 records, each representing a separate fire incident. The millions of fire incidents that have been entered make NFIRS the world's largest fire incident database.

National Center for Health Statistics. Injury and mortality risk estimates are based on 2001 data from the Centers for Disease Control and Prevention's (CDC) NCHS Mortality database. The Mortality data base is a compendium of deaths in the United States. Data are extracted from death certificates, and the cause of death is coded in compliance with the International Classification of Diseases (ICD). Only a few of these codes apply to fire or fire related mortalities. Used in conjunction with NFIRS, these data sets provide answers to questions about the relationship of casualties to fire in the United States. NFIRS data answer questions about the incident and the victim's relationship to the fire. NCHS data yield specific numbers of fatalities by age group and the ethnicity of the victims.

In addition to the mortality data, the CDC and NCHS conduct national surveys on health related issues that provide valuable data and insight into the health issues and subsequent risks for the older population.

Census Bureau. The U.S. Census Bureau population estimates, projections, and other demographic surveys and reports provide valuable profiles of older Americans not just by age, but also by how and where they live.

Adjusted Percentages in Fire Data

In making national estimates of the fire problem, unknown or undetermined data in the NFIRS data base are not ignored. Unknown data occur when the information in nonrequired data collection fields in NFIRS is not provided (left blank), the coding is invalid, or the information is entered as "undetermined." This report provides an "adjusted" percentage that is computed using only those incidents for which the valid information was provided for the data item being analyzed. In effect, this distributes the unknown responses in the same proportion as the known responses for the data item.

Spurred by current day advancements in medical science, the population of Americans 65 years and older is at its highest point in history, and the U.S. Census Bureau projects that the increases will con tinue. To comprehend and address the risk of fire to older Americans, it is essential to first understand the demographics of the senior population.

Increases in the 65 and Older Population: The Graying of America

The American population is aging. Although currently growing at a moderate pace, the older U.S. pop ulation is projected to increase at a faster rate in the near future. According to the U.S. Census Bureau, adults aged 65 years and older were 4 percent of the population in 1900. By 2000, older adults com prised 12 percent of the U.S. population and numbered 35 million, an all time high. [Refs. 1 and 2] In what has become known as the "graying of America," we are fast becoming a nation of the elderly. Over the past century, the number of persons under the age of 65 tripled while those over age 65 increased elevenfold. [Ref. 3] The U.S. Census Bureau estimates that between now and the year 2050, the demographic landscape of America will become even grayer. The senior population is expected to more than double, reaching nearly 87 million, or approximately 21 percent of the projected 2050 American population (Figure 1). [Refs. 1 and 4]

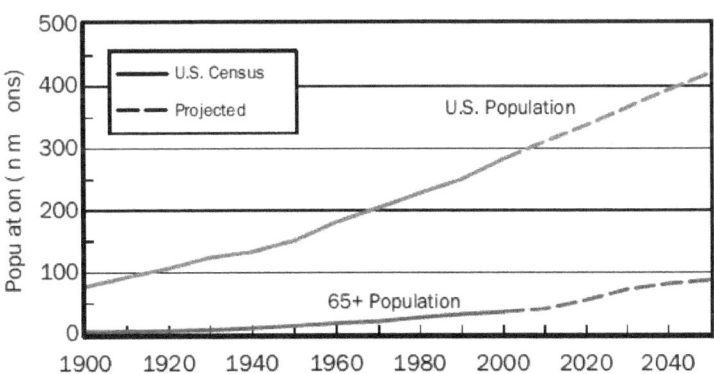

Sources: Population estimates 2010 2050 from file natprojtab02a.xls in
Reference 1; population estimates 1900 2000 from Appendix A,
Detailed Tables, Tables 1 and 5 in Reference 4.

Figure 1. U.S. Population Growth

Most of this growth is expected to occur between the years 2010 and 2030, when the baby boom generation enters the 65 and older population group. As of 2003, the 77 million people born between 1946 and 1964 constitute 30 percent of the entire U.S. population, average 57 years of age, and will enter their mid 60s between 2010 and 2030. [Ref. 5]

Global Increases and Trends
Internationally, senior populations are increasing at an even faster rate than in the United States. The population of people 65 years and older increases worldwide by 795,000 every month, with the largest increases occurring in developing countries such as Singapore, Malaysia, Colombia, Costa Rica, and the Philippines. Developing countries are projected to have the highest increases in the older population in the next 30 years. The United States is projected to have the 29th highest percentage increase in the older population during this period. [Ref. 6]

Profile of the Older Adult

The demographic characteristics of older adults reveal both commonalities and differences. Age, gender, race, place of residence, and marital status are factors that may influence the risk of having a fire and the subsequent risk for fire fatality and injury.

AGE

Roughly half of older Americans were between ages 65 and 74, one third were between ages 75 and 84, and one sixth were over 85 in 2000 (Figure 2). [Ref. 7] The "oldest old" (aged 85 and older) is the most rapidly growing age group among older adults. Between 1960 and 2000, their numbers rose by 356 percent. In contrast, the senior population in general rose by 111 percent, and the entire U.S. population grew by 57 percent. By 2050, the oldest old will number over 20 million, or 24 percent of the 65 and over population and 5 percent of the total U.S. population. [Ref. 1]

In 2002, more than 2 million people in this country celebrated their 65th birthday. Medical and health care advances have extended life expectancy. In particular, geriatric medicine has contributed significantly to lengthening the average lifespan of Americans today. On average, those who reach age 65 can expect to live an additional 18 years. [Ref. 8]

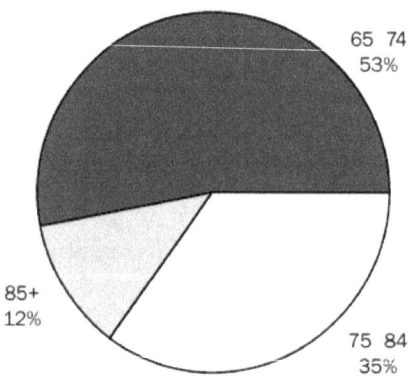

Figure 2. Age Distribution of Older Adults in 2000

Table 1 shows the steady climb in the average life expectancy over the past century—an increase of 63 percent. At the turn of the century, infectious disease was the leading cause of death and was largely responsible for limiting the average lifespan to about 50 years. [Ref. 10] Thanks to widespread immu nizations and medical research, we have eradicated all but a few of the most deadly killers of that cen tury. In the latter half of the 20th century, heart disease and cancer replaced infectious disease as the leading causes of death in the United States. Further, as the average age of a population changes, so does the picture of health for that population. As people live longer, there will be a substantial increase in the numbers who face dependency as the result of chronic illness or disability.

Table 1. Average Life Expectancy (1900 to 2002)

Year	Total Population	Male	Female
1900	47.3	46.3	48.3
1910	50.0	48.4	51.8
1920	54.1	53.6	54.6
1930	59.7	58.1	61.6
1940	62.9	60.8	65.2
1950	68.2	65.6	71.1
1960	69.7	66.6	73.1
1970	70.8	67.1	74.7
1980	73.7	70.0	77.4
1990	75.4	71.8	78.8
2000	77.0	74.3	79.7
2002	77.3	74.5	79.9

Source: Reference 9.

GENDER

More women than men populate the older age brackets, and the proportion of women to men increases with age (Table 2). In 2001, older American women numbered 21 million and accounted for approxi mately 59 percent of all older Americans. Fifty five percent of older Americans aged 65 to 74 were women; for those 85 and older, 70 percent were women. [Ref. 11]

RACE

In 2003, Americans 65 years and older were predominately non Hispanic whites (83 percent). Eight percent were African American, 6 percent were Hispanic of any race, 3 percent were Asian or Pacific Islander, and less than 1 percent were American Indian, Native Alaskan, or other race. [Ref. 8]

By 2050, experts predict that the U.S. population will become more diverse with the older population consisting of 61 percent non Hispanic white, 18 percent Hispanic, 12 percent black, and 8 percent Asian (Table 3). [Ref. 7]

Table 2. Resident Population Estimates by Age
and Gender in 2001

Gender	2001 Population	Percent
All Older Adults (Age 65+)		
Male	14,619,070	41
Female	20,734,196	59
Total	35,353,266	
Age 65–74		
Male	8,301,935	45
Female	10,020,545	55
Total	18,322,480	
Age 75–84		
Male	4,996,556	40
Female	7,585,928	60
Total	12,582,484	
Age 85+		
Male	1,320,579	30
Female	3,127,723	70
Total	4,448,302	

Source: Reference 11.

Table 3. Population Aged 65 and Over by Race and Hispanic Origin –
2003 and 2050 (projected)

Race/Ethnicity	Percent Population	
	2003	2050 (Projected)
Non Hispanic White	83	61
Non Hispanic African American	8	12
Non Hispanic Asian	3	8
Other Race, Non Hispanic	1	3
Hispanic, Any Race	6	18

Source: U.S. Census Bureau, Population Estimates and Projections, 2004 as shown in Reference 7.

GEOGRAPHY

More than half of Americans 65 years and older live in the nine most populous States: California, Florida, New York, Texas, Pennsylvania, Ohio, Illinois, Michigan, and New Jersey. These same States account for over half of the U.S. population. The South and Midwest have the largest number of senior residents as a percentage of the region's overall population. [Ref. 2]

Table 4 shows the 10 cities with the highest percentage of older Americans in relation to the size of the total population; 6 are in Florida.

Table 4. Cities With the Highest Percent of Aged 65 and Over Population in 2000

City	Total Population	65+ Population	% Overall Population
Clearwater, Florida	108,787	23,357	21.5
Cape Coral, Florida	102,286	20,020	19.6
Honolulu, Hawaii	371,657	66,257	17.8
St. Petersburg, Florida	248,232	43,173	17.4
Hollywood, Florida	139,357	24,159	17.3
Warren, Michigan	138,247	23,871	17.3
Miami, Florida	362,470	61,768	17.0
Livonia, Michigan	100,545	16,988	16.9
Scottsdale, Arizona	202,705	33,884	16.7
Hialeah, Florida	226,419	37,679	16.6

Source: Reference 2, Table 5.

MARITAL STATUS

According to the 2000 census data, 73 percent of men and 41 percent of women 65 years and older are married. Among all older adults, 46 percent of women and 14 percent of men have no living spouse, and 10 percent of all older adults are divorced or separated. [Ref. 8]

As people age, they undergo physical and cognitive changes that can impair their abilities to prevent or respond to fires and situations in which fire is likely. In addition, many older adults suffer mobility impairments (disabilities), which make their ability to successfully escape fires more difficult. Some substance induced impairments, such as those caused by alcohol consumption or the side effects of prescription medication, can result in a wide range of impairments that increase older adults' risks of fire, fire fatality, and fire injury.

When evaluating the fire risk to older adults, it is important to remember that no single risk factor exists in a vacuum. Often, one physical and cognitive change does not stand alone. Someone could have poor vision, poor hearing, and be prescribed impairing medications—all at the same time.

As life expectancies for older Americans rise—a person 65 years of age is expected to live 18 more years [Ref. 7]—the impact of physical and cognitive impairments will have an increasing impact on the risk of fires and the resulting injuries and deaths to older Americans.

Sensory and Cognitive Impairments in Older Adults

Sensory changes, such as an impaired sense of smell, touch, sight, or hearing, can increase the risks of fire to older adults. When several senses are weakened in a person, the risk to an individual is com pounded. Cognitive changes, or changes in mental functioning, constitute additional fire risk factors. A decrease in mental cognition can often be worse than a decrease in physical adeptness. Individuals who suffer from reduced mental faculties often do not realize they are in any danger and can some times even engage in risky behavior.

SMELL

A person's sense of smell can be a good fire detector, and certainly a welcome complement to a work ing electronic smoke alarm. Recent research, however, indicates that the sense of smell is not effective at detecting smoke or fires during sleep [Ref. 12], although during waking hours the smell of "some thing burning" often leads to a person's discovery of a fire, a potential fire, or a stove or oven left on after cooking concludes. The sense of smell diminishes as people age. In older adults, this loss can occur in tandem with other physical and mental problems that are themselves fire and fire injury risks.

A study by researchers at San Diego State University revealed that a majority of older adults have a weaker sense of smell than younger adults. The study also drew a correlation between this diminished sense of smell and such mental faculties as memory and reasoning ability. [Ref. 13] It has been widely observed that smell and taste losses have a negative impact on quality of life. [Ref. 14]

Smell impairments range from a generally diminished sense of smell to complete inability to detect odors at all, called anosmia. Some people experience an altered sense of smell, called dysosmia. Common causes of smell impairments are [Ref. 14]

- nasal and sinus disease;
- upper respiratory infection;
- head trauma;
- cigarette smoking; and
- neurodegenerative disease (e.g., Alzheimer's or Parkinson's disease).

Whatever the individual cause, a decreased sense of smell is a likely impairment as a person's body ages. Recent research indicates that by age 60, 30 percent of individuals have some olfactory impairment. By age 80, this percentage has risen to over 60 percent. [Ref. 15]

TOUCH

The simple act of feeling whether something is hot or not can be a fire preventative measure. Touch can indicate whether an appliance is overheating or a doorknob is scorching, and it can trigger a quick reaction to prevent a fire. The acuteness of this sense can be measured by the ability of a person to accurately gauge the difference between cool, cold, neutral, warm, and hot and to feel and hold things normally.

Many older adults experience a decreased sense of touch, a problem that manifests itself in several forms. Individuals with a weakened or decreased sense of touch are more likely to have difficulty performing daily tasks such as unplugging electrical cords. [Ref. 16] Touch problems associated with aging, thinning skin can lead to a higher risk for burn injuries and increase the severity of burn injuries.

Researchers attribute skin changes that decrease the sense of touch in older adults to several factors, including [Ref. 17]

- the natural thinning of the skin's outer layer, the epidermis;
- cumulative effects of sun exposure;
- blood vessels becoming more fragile;
- changes in connective tissue which, in turn, reduce skin strength and elasticity;
- the side effects of medications; and
- dehydration.

In addition to other risks, older skin can take up to four times longer to heal than the skin of a younger person, making the risk of fire and fire injury an even more critical issue for the older adult. [Ref. 17]

VISION

With 18 percent of older Americans currently suffering from eye impairments, vision loss is one of the more severe fire and fire casualty risk factors among older adults. [Ref. 7] The inability to see fires, to notice fire causes such as falling candles or cigarettes that have tumbled from bedside ashtrays, and to locate doors and escape routes are all obvious risks of fire and fire injury. Older adults with diminished sight are more likely to stumble and fall when in a hurry, impeding their ability to escape a fire.

The most typical changes to the eye caused by aging are loss of focus, declining color sensitivity, and a need for more light. [Ref. 18] Presbyopia, the inability to focus on close objects as a result of age related changes to the eye and lens, is often solved by simply wearing glasses. But there may be a limit to vision correction from glasses alone. Also, finding and wearing the glasses may be problematic for older adults, especially when awakened at night in an emergency situation.

The most severe conditions associated with impaired vision are more difficult to solve, and many older adults are forced to adapt to these problems. Such conditions include [Ref. 19]

- macular degeneration—the gradual loss of detailed vision, sudden loss of central vision, and need for more light to see;
- glaucoma—the subtle loss of contrast and loss of peripheral vision;
- cataracts—cloudy vision, difficulty discerning colors and increased sensitivity to glare; and
- diabetic retinopathy—blurred vision and near vision distortion.

A publication by Lighthouse International, an organization that studies vision impairment, states that seniors can reduce the risks associated with vision impairment by adapting their homes and lifestyles to accommodate their diminished sight. [Ref. 20]

HEARING

The ability to hear the blare of a smoke alarm and other warning sounds of fire is something many people with excellent hearing take for granted. For the approximately one third of older Americans who have hearing impairments, the inability to hear such vital warnings represents a substantial fire risk factor.

Hearing impairments afflict 37 percent of all Americans 65 years and older. This percentage increases with age: hearing problems were reported in 30 percent of adults aged 65 to 74; 42 percent in adults aged 75 to 84; and 60 percent in those aged 85 and over. Hearing loss affects more men than women. [Ref. 7] The CDC reports that older adults are less likely to go to the doctor to treat hearing problems than for vision problems. [Ref. 21]

Hearing loss in older adults is attributed to exposure to loud noises over long periods of time, the effects of smoking, a history of middle ear infections, and to certain chemicals. [Ref. 21]

Preliminary research and anecdotal evidence also suggest that older Americans have difficulty responding to smoke alarms while sleeping. In 2003, the U.S. Consumer Product Safety Commission launched a 2 year study to investigate smoke alarm waking effectiveness among children and older adults after research indicated that those groups had difficulty hearing or responding to the sound from smoke alarms. [Ref. 22]

MEMORY IMPAIRMENTS, DEMENTIA, AND ALZHEIMER'S DISEASE

Among the older population, 15 percent of men and 11 percent of women suffered from some form of moderate to severe memory impairment in 2002. [Ref. 7]

Nationally, 4.5 million people are inflicted with Alzheimer's disease. According to the Alzheimer's Association, one in ten people 65 years and older have the disease. In the oldest age group, those 85 years and over, almost half are affected by Alzheimer's. The association estimates that 16 million Americans will have the disease by the year 2050 if no cure is found. The average life expectancy after an Alzheimer's diagnosis is 8 years. [Ref. 23]

Because memory impairments directly affect reasoning and basic memory, they are substantial fire risk factors for older adults. With dementia and Alzheimer's disease, such mental impairments make out of the ordinary behaviors possible, including dangerous actions and fire risky behaviors. Accidents, falls, and contact with dangerous substances are more prevalent among dementia patients, and for such patients, living quarters should be modified to remove anything within reach that could pose a potential fire risk. [Ref. 24]

DEPRESSION

Depression, which affects about 18 percent of older women and 11 percent of older men, poses a more complex risk. [Ref. 7] The wide ranging effects of depression make it more difficult to analyze in relation to fire risk. Common symptoms of depression that could impact that risk include fatigue or loss of energy, recurrent thoughts of fatality or suicide, difficulty paying attention or making decisions, and confusion and avoidance. [Ref. 25]

According to the National Health and Retirement Survey, older women are more likely to experience depression than older men, and the propensity for depression in both men and women increases with age. Among people 85 years and older, 20 percent experienced signs of depression, compared with an average of 13 percent of people between the ages of 65 and 69 who experienced the same signs. [Ref. 7]

Disabilities and Mobility Impairments

In 1999, approximately 20 percent of older Americans were classified as "chronically disabled." [Ref. 7] Many others, as many as half of the older population, suffer from some form of disability or mobility impairment. [Ref. 8] Thus, the ability of older adults to react to situations, respond to fires, and escape burning structures is hampered when their movement is slowed or impaired.

DISABILITIES AND ACTIVITY LIMITATIONS

Mobility impairments range from general slowness to complete disability. The U.S. Department of Health and Human Services evaluates disability and health by individual performance in two catego ries, activities of daily living (ADL) and instrumental activities of daily living (IADL). ADLs include eating, dressing, getting in and out of bed, getting around inside, bathing, and toileting. IADLs are considered more detailed tasks, such as heavy and light housework, laundry, preparing meals, shop ping for groceries, getting around outside, traveling, managing money, and using a telephone. [Ref. 7]

Figure 3 shows that disabilities of all types increase with age. [Ref. 8]

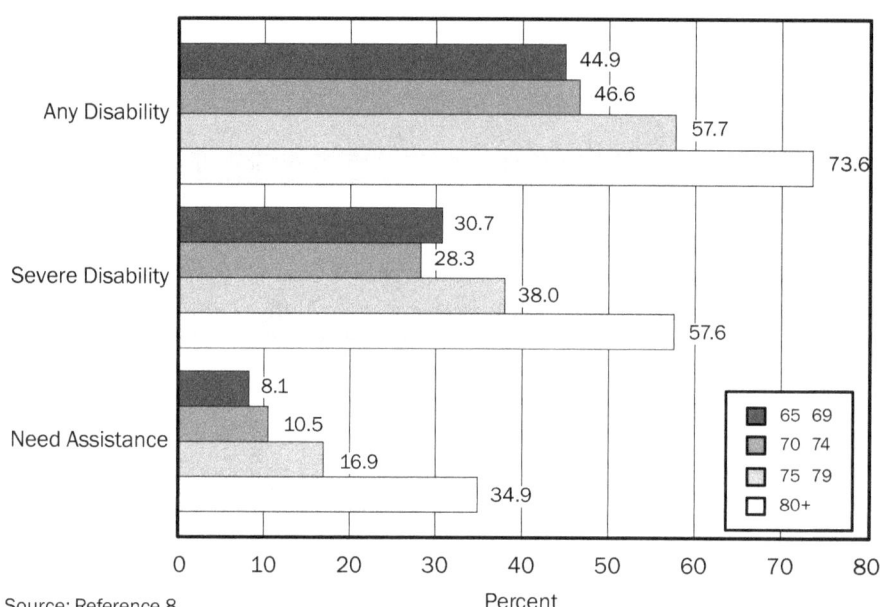

Source: Reference 8.

Figure 3. Percentage of People Aged 65 and Older With a Disability, 1997

FALLS

Falls have a serious impact on the health of older Americans, and thus a serious impact on the fire risks facing seniors. Approximately one third of adults 65 years and older fall each year, with 10 percent incurring a serious injury. [Ref. 26] Within 6 months, two thirds of those who fall will fall again. [Ref. 27] Among older Americans, falls are a leading cause of serious injuries and injury caused deaths. [Ref. 27] In 2000, falls accounted for $16 billion in direct medical and long term care costs among adults 65 years and older. [Ref. 28]

Fifty five percent of all fall injuries among persons 65 years and older took place inside the home. An additonal 23 percent occurred outside near the home. [Ref. 28]

Twenty eight percent of fall injuries among adults 65 years and older resulted in ADL or IADL limita tions, compared with a 16 percent ADL or IADL impact on adults aged 35 to 64. [Ref. 28] Falls cause

87 percent of fractures treated in emergency departments and are the second leading cause of spinal cord and brain injuries among older Americans. [Ref. 27]

According to the National Center for Injury Prevention and Control, falls are caused by personal factors such as muscle weaknesses, balance problems, limited vision, and taking certain medications such as tranquilizers or antidepressants. Environmental factors, such as home clutter, loose rugs, poor lighting, and the lack of grab bars in bathrooms, also contribute to the cause of falls. [Ref. 27]

Of older households with a physically limited person, slightly less than half (49 percent) made any modification to increase the home's safety or accessibility, making fire exit strategies an additional concern and compounding other risk factors. [Ref. 28]

Alcohol and Prescription Drugs

Prescription drugs and alcoholic beverages, alone and especially when combined, are fire risk factors similar to cognitive impairments. The decreased alertness and impaired mental lucidity associated with prescription drugs, alcohol, and the combination of both are clear fire risks.

ALCOHOL ABUSE

For older adults with or without cognitive or physical impairments, the over consumption of alcoholic beverages slows reactions, impairs coordination and alertness, causes drowsiness or sleepiness, and can cause forgetfulness and confusion. [Ref. 29] For the older adults with age related impairments or disabilities, alcohol can be a substantial fire risk, as alcohol abuse can render a person incapacitated. It also compounds other risk factors.

According to the National Institute on Alcohol Abuse and Alcoholism, alcohol has a more pronounced effect on older adults because of decreased body water content. Thus, it takes less time for an older adult to feel the effects of alcohol or become inebriated than a younger person. Moreover, many years of heavy drinking can damage the brain, central nervous system, and bodily organs including the liver, heart, kidneys, and stomach. [Ref. 29] Alcohol consumption increases the likelihood of falls and hip fractures, again compounding the fire risk for older adults.

As shown in Table 5, the CDC's National Center for Chronic Disease Prevention & Health Promotion reports that older adults drink alcohol on more days every month than younger adults. Although an older adult will generally consume a smaller quantity of alcohol in one sitting than a younger adult, 6 to 11 percent of the elderly exhibit symptoms of alcoholism. In nursing homes, drinking problems can reach as high as 49 percent of the home's population. [Ref. 30]

Table 5. Adult Population Reporting Alcohol
Consumption 21 to 31 Days per Month

Age Group	Percent
18 24	6.5
25 34	6.5
35 44	9.5
45 54	13.2
55 64	17.3
65+	24.5

Source: Reference 31.

PRESCRIPTION DRUGS

Prescription drugs are necessities for seniors suffering from a wide range of ailments, including heart disease, hypertension, cancer, stroke, lower respiratory diseases, and diabetes. [Ref. 7] Seniors account for 30 percent of all prescription drug users in the Nation—older prescription drug users now take more than four different prescription drugs along with two over the counter medications. [Ref. 32] The average older adult spent $955 on drugs in 2002. [Ref. 8] Usage of prescription drugs can lead to side effects that are definite fire risk factors, including confusion, memory loss, disorientation, blurred vision, higher risk and rate of falls, and dizziness. [Ref. 33] Sometimes these side effects are mistakenly diagnosed as dementia or depression, for which additional drugs are then prescribed.

Prescription drugs are not always used correctly by older adults, thus adding to behaviors that increase fire risk. Due to the high cost of prescription drugs, 25 percent of older adults surveyed in an eight State study skipped doses of their prescription drugs or did not fill a prescription at all. This can lead to physical instability and can make it more difficult for older adults to adapt to the side effects of their prescription drugs. [Ref. 34]

It is estimated that 15 percent of hospitalizations of older adults are caused by adverse reactions to med ication. [Ref. 35] According to the National Pharmaceutical Council, the risk of an adverse prescription drug reaction (ADR) in elderly individuals increases as the number of medications those individuals take increases. [Ref. 36] Such reactions can cause additional impairments, which only increase the risk of fire, fire injury, and fire fatality.

MIXING ALCOHOL AND DRUGS

Taken alone, alcohol and prescription drugs each increase the risk for fire and fire injury. Taken together, the hazards multiply. Heavy drinking can activate the bodily enzymes that break down toxic substances, which can impair the effectiveness of prescription drugs or cause immediate ADRs. The regular side effects of the drugs remain, while their intended result—to treat a condition—is not achieved. [Ref. 30] More importantly, the interactions between alcohol and drugs often heighten the side effects of both substances, especially drowsiness, making the fire risk for drug and alcohol mixing sub stantial. [Ref. 37]

Economic and Social Factors

Lower income and impoverished older adults often cannot complete necessary home repairs, buy medications essential for maintaining their physical health, or replace aging electrical appliances, placing them at higher risk for fire. Additionally, educational or social factors can interfere with an older adult's ability to understand the details regarding fire prevention and safety.

THE ECONOMIC STATE OF OLDER ADULTS

Of Americans 65 years and older, 14.4 percent participated in the workforce in 2004, constituting 3.3 percent of the overall U.S. labor force. [Ref. 38] Of these older workers, more older men (55.7 percent) worked than did older women (44.3 percent).

In 2002, about 32 percent of all adults 65 years and older reported incomes of less than $9,999 per year. Figure 4 shows the economic breakdown of individual older Americans. [Ref. 8]

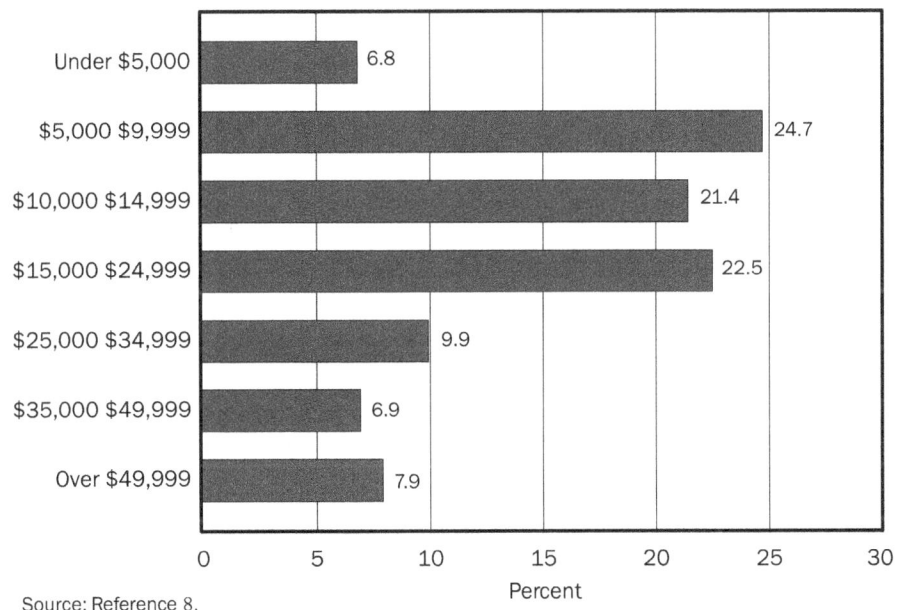

Source: Reference 8.

Figure 4. Annual Income for Aged 65 and Older Adults in 2002

Since the mid 1960s, the number of older Americans living in poverty has decreased substantially. In 1966, Americans 65 years and older represented the largest impoverished age group in the Nation, with 28.5 percent of that age group living in poverty. In 2002, 10.4 percent of Americans 65 years and older lived in poverty. [Refs. 7 and 39] Older women (12.4 percent) were more likely to live in poverty than older men (7.7 percent), and older African Americans (23.8 percent) and older Hispanics of any race (21.4 percent) were far more likely to live in poverty than older non Hispanic whites (8.3 per cent). [Ref. 7] The most impoverished groups among seniors are African American and Hispanic women. [Ref. 7]

INCOME AND HEALTH

As previously discussed, overall health and physical and mental impairments are key fire risks for older adults. Often these impairments result from lack of medical care by the elderly—or soon to be elderly—due to financial concerns. There is a correlation between socioeconomic status and the health of older adults, meaning that seniors living in poverty, who often have not received health care throughout their lives, have worse health than individuals who can afford and do seek long term care. According to a study by RAND Corporation's Center for the Study of Aging, lifelong health care trends in individual older adults—affected by their lifelong socioeconomic positions—most directly affect the status of their health in their older years. [Ref. 40]

Data from the first 6 months of 2004 show that 12.6 percent of people aged 45 to 64 (tomorrow's older adults) did not have any health insurance, and 7.1 percent of all people aged 18 to 64 failed to obtain preventive or needed medical care during the course of a calendar year because of a financial barrier. [Ref. 41] Without health insurance, these individuals often enter their senior years in poor health.

Some older adults can simply not afford medical care: 2.5 percent of adults 65 years and older failed to obtain needed medical care in the previous 12 months because of cost. This percentage has been essentially constant over recent years. Women are more likely than men to lack access to medical care due to cost. [Ref. 41]

Populations with lower incomes and higher rates of poverty tend to have worse health than their con temporaries. Older African Americans and older Hispanics are less likely than older whites to rate their health as excellent or good. [Ref. 8]

LIVING ARRANGEMENTS

Economic and social factors affect older adults' living arrangements. Finances can dictate where they live and with whom, and if they are able to afford arrangements that include health care and living assistance.

Seniors overwhelmingly live in homes that they own. In 2003, there were 21.6 million elderly house holds, according to the American Housing Survey (AHS). Of these elderly households (AHS defines elderly households as those headed by an individual 65 years or older), 17.3 million elderly house holds—80.2 percent—were owner occupied. When rental and owner occupied housing types are considered together, the majority of elderly households—69.2 percent—are single family, detached homes. Of the other housing unit types occupied by Americans 65 years or older, 23.9 percent are multifamily units such as duplexes, apartments, or condominiums. Another 6.9 percent of elderly housing units are mobile homes or trailers. [Ref. 42]

Of all Americans 65 years and older, 73 percent live with a spouse, 5 percent live with other relatives, 3 percent live with non relatives, and 19 percent live alone. [Ref. 7] Only 4.5 percent of the senior popu lation lived in nursing homes in 2000, down from 5.1 percent in 1990. [Ref. 2] However, 2000 Census data indicate that as individuals age they are increasingly more likely to live in nursing homes. In 2000, for example, 18.2 percent of persons 85 years or older lived in nursing homes compared with only 1.1 percent of those aged 65 to 74. [Ref. 2]

Where older adults live, and with whom, affects their risk of fire and fire fatality and injury. Older adults who own their own homes or live with relatives must recognize their mobility limitations and cognitive impairments and install safety devices in order to reduce the risk of fire fatality and injury. They should also replace old and outdated appliances and electrical devices that could pose fire risks. Living with family members who can respond more quickly to fire risks or in long term care facilities could be safer than living alone.

The fire risks inherent in long term care and home health care facilities are discussed in more detail in the following two chapters of this report.

INCOME AND FIRE DETECTION AND PREVENTION

Sixty nine percent of seniors claim that they could not afford to make necessary improvements to their home, including installing wheelchair ramps, providing grab bars in bathrooms, and making other improvements that enhance the livability of the home for older adults. [Ref. 43] Food, housing, medical costs, and other daily living expenses are higher priority items to seniors living on small, fixed incomes. Therefore, income streams to older adults can play an important role in their safety during a fire emergency. Upgrading smoke alarms, repairing or replacing potential fire hazard appliances, and taking other fire safety measures may take a financial backseat.

Older adults also may not have the same level of protection as younger people in housing environ ments. Apartment complexes must comply with rigorous fire codes and laws and make repairs and modifications to observe local ordinances; they are inspected to ensure compliance. Individual homes, even if required to comply with related laws, are rarely inspected. Older adults account for nearly one fourth of all homeowners in the Nation. [Ref. 42] Low income seniors may choose to live in their own homes, but, because of high costs, never replace dangerous, outdated, and flammable housing materi als that can increase fire risk.

PUBLIC EDUCATION AND SOCIAL INTERACTION

Understanding the risks for fire and fire injury is one of the first steps towards preventing fires. Unfor tunately, many lower income older adults may not be reached or affected by public information pro grams aimed at educating them about such risks. Older adults with little social interaction (such as involvement at a senior citizens' center) may also have no exposure to fire prevention programs and thus may not understand their fire risks. Today's seniors were raised in a time where fire prevention was not as highly emphasized in schools as it is today.

Many researchers note that a strong social structure has a positive effect on quality of life and may con tribute to good health. Of all older adults with difficulties performing two or more ADLs, however, 20 percent say they do not have regular interaction with family and 36 percent do not have regular interactions with friends. [Ref. 43] According to the American Association of Retired Persons (AARP),

77 percent of older adults and future older adults view having a senior citizens' center in their neigh borhood as important for their health and happiness. [Ref. 44] Such facilities may not be available in impoverished areas where low income seniors reside. Frequently, these centers run programs or dis tribute literature on fire safety tailored specifically to older adults. Other services that seniors view as important, such as a personal care service, light home repair service, contractor service, and nutrition programs, also may not be available in communities where many low income seniors reside.

Chapter 5 - LONG-TERM CARE FACILITIES

In 2003, 31 residents died in nursing home fires in Hartford, Connecticut, and Nashville, Tennessee. The incidents raised awareness of fire risks in long term care facilities, and emphasized the need for more research and education regarding such risks. [Ref. 45]

Long term care facilities, which include nursing homes and assisted living facilities, provide many opportunities and amenities for their patients. Convenient services, health monitoring, assistance with activities of daily living, and onsite staff can offer comfort, peace of mind, and needed attention. These facilities, however, present unique fire risks that are often different and more complex than the risks posed in residential structures. With a high percentage of nursing and assisted living facility patients disabled, the presence of highly flammable gases and liquids, and structure architecture that can make evacuation difficult, the fire risks in nursing homes and assisted living facilities pose daunting challenges to the fire service in the event of fires.

Nursing Homes

Nursing homes are facilities with three or more beds that routinely provide nursing care services. According to the most recent (1999) National Nursing Home Survey by the CDC, nearly 2 million people live in nursing homes and more than 1 million of them are 65 years of age and older. [Ref. 46] Medicare and Medicaid payments for care in nursing homes equaled $64 billion in 2002. [Ref. 45] According to the CDC:[1]

- There are 18,000 nursing homes in the United States.
- Forty percent have between 100 and 199 beds, 39 percent have 50 to 99 beds, 12 percent have fewer than 50 beds, and 8 percent have more than 200 beds.
- More than 80 percent of nursing homes are certified by both Medicare and Medicaid.
- Two thirds of nursing homes are in the Midwest and South.
- Two thirds of nursing homes are run by for profit corporations.
- Nursing homes employ about 1.5 million full time employees.

Over 90 percent of nursing home residents are 65 years or older. Of older adults who are nursing home residents, 41.7 percent are 85 years and older, 39.8 percent are between 75 and 84 years old, and 18.6 percent are between 65 and 74 years old.

---

[1]The remainder of this section on nursing homes is drawn from information in Reference 46.

Residents of nursing homes are predominately female and white. Of all residents, 72 percent are women and 28 percent are men. Eighty six percent of residents are white and 11 percent are African Americans. This statistic is consistent with the demographics of older Americans.

Nursing home populations have a wide range of health conditions; the most prevalent conditions are

- mental disorders, including senile dementia, organic brain syndrome, mental retardation, and other diseases;

- nervous system and sensory diseases, including Alzheimer's disease, Parkinson's disease, multiple sclerosis, and paralytic syndromes; and

- circulatory system diseases, including hypertension and heart disease.

Of all nursing home residents, 75 percent require assistance with three or more ADLs. Figure 5 shows the percentage of residents requiring daily living assistance.

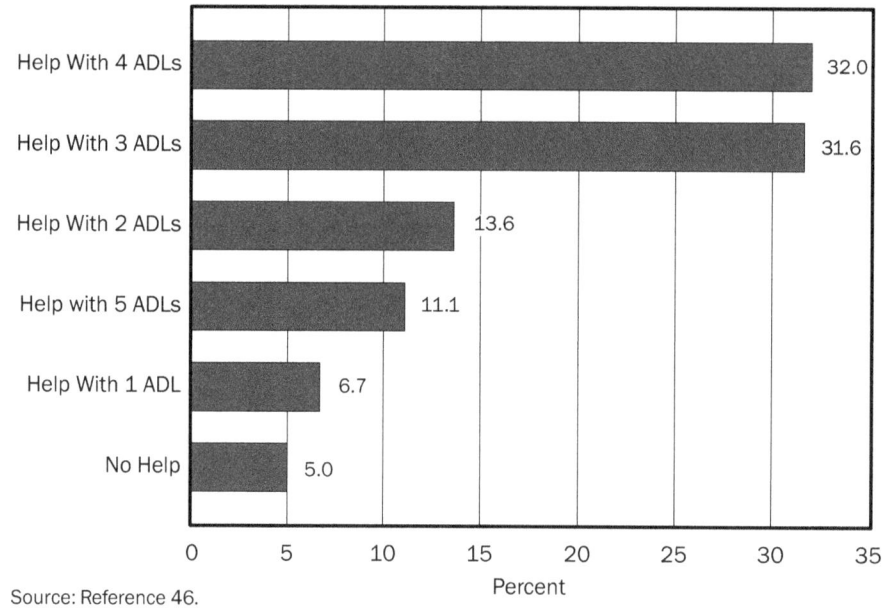

Source: Reference 46.

Figure 5. Nursing Home Residents Needing Assistance
With Activities of Daily Living

The CDC reports that most of the residents who need assistance require help with bathing (94 per cent), dressing (87 percent), toileting (56 percent), and eating (47 percent). Most residents wear eye glasses and use a wheelchair or a walker (Figure 6).

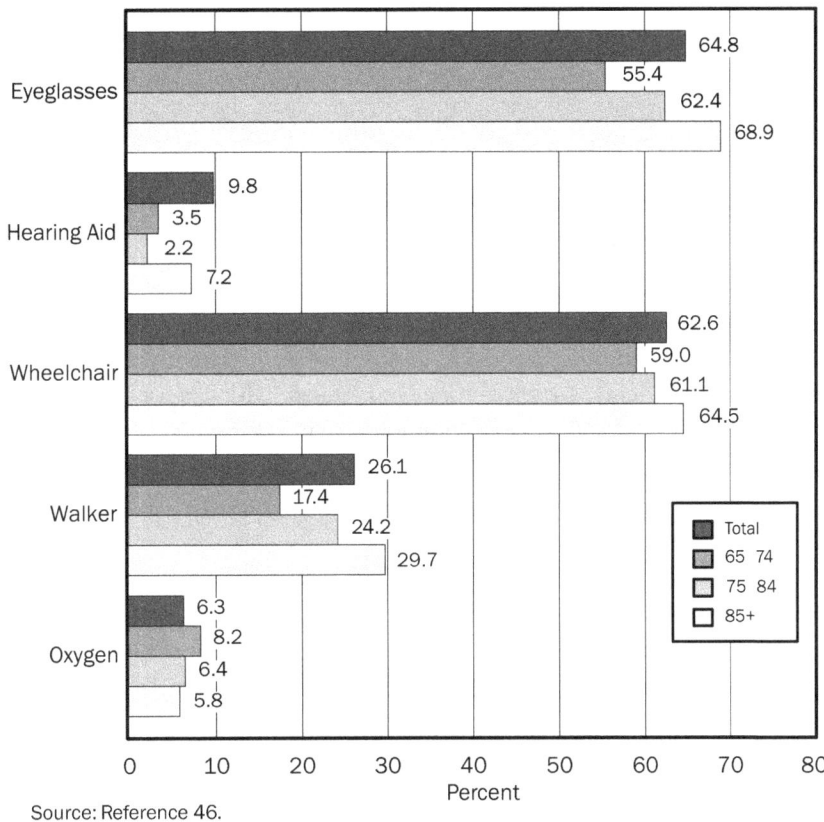

Source: Reference 46.

Figure 6. Nursing Home Residents Requiring Aid

Assisted Living Facilities

According to AARP and the insurance provider MetLife, the term *assisted living facility* is vague and encom passes many different types of housing situations for older adults. In fact, different States define the term with varying characteristics. Most definitions include assistance with ADLs, meal preparation, 24 hour supervision, and housekeeping. AARP has published assisted living facility philosophies that are often used as the standard for evaluating such facilities. Under this philosophy, assisted living facili ties should [Ref. 47]

- maximize residents' personal dignity, autonomy, independence, and privacy;
- minimize the need to move when a resident's need changes; and
- meet residents' needs, both scheduled and unscheduled.

According to the National Center for Assisted Living and the Centers for Medicare and Medicaid Ser vices: [Refs. 48 and 49]

- In 2000, there were 32,886 licensed assisted living facilities in the United States providing 795,391 beds for older adults nationwide.

- From 1992 to 1998, the number of seniors living in facilities increased by 56 percent from 267,000 to 417,000.

- No federal quality standards exist for assisted living facilities.

- The average yearly cost of an assisted living facility in 1999 was $15,332, often a higher out of pocket cost than a nursing home.

The average age of the assisted living facility resident is 85. Most residents (79 percent) are female. The assisted living facility population in the United States is 99 percent white. [Ref. 47]

Of assisted living facility residents, 81 percent need assistance with one or more ADLs. The average resident needs help with two ADLs. The most common activities for which they require assistance are bathing (30 percent), dressing (24 percent), and toileting (19 percent). The IADLs for which residents require the most assistance are meal preparation (80 percent), housework (73 percent), and traveling (66 percent). [Ref. 48]

According to the AARP, 52 percent of all assisted living facility residents have a cognitive impairment of some type. [Ref. 47]

Fire Risks in Long-Term Care Facilities

Long term care facilities are among the most regulated and inspected when it comes to fire safety. Most nursing homes, for example, are regulated by codes based on National Fire Protection Association (NFPA) standards that take into account the construction and design of facilities; provision for fire detection, alarm, and extinguishment; and fire prevention policies. Also scrutinized are provisions for plans that test staff response to fires and other emergencies.

Despite these stringent regulations, fire risks still exist and some nursing homes are exempt from the requirements. According to the U.S. Government Accountability Office (GAO):

> ...the degree to which the standards rely on staff to protect and evacuate residents may be unrealistic. Moreover, many unsprinklered [nursing] homes are not required to meet all fed eral fire safety standards if they obtain a waiver or are able to demonstrate that compensating features offer an equivalent level of fire safety. However, some of these exemptions raise a concern about whether resident safety was adequately considered. [Ref. 45]

Accordingly, it is important to consider the fire risks in long term care facilities, which include the health of residents, building systems, flammability and combustibility of medical materials, type of architecture, and the availability of sprinklers and smoke alarms.

HEALTH OF RESIDENTS

As discussed in Chapter 4, impaired health is a risk factor for fire fatality. Residents of long term care facilities have a higher rate of illness, disability, mental illness, and medication needs and require more assistance than the general population. These health issues can increase the risk of fire and are key factors in fire fatalities and injuries. In a fire situation, it may be difficult to evacuate a long term care facility filled with incapacitated residents before fire takes its toll. The propensity for injury and fatality is clearly higher than in a complex where all inhabitants have full mobility.

Strict policies in nursing homes may alleviate some fire risk factors. Nursing home residents are often not permitted to smoke, light candles, or possess dangerous materials (such as flammable clothing and blankets and appliances with frayed wires). They are also monitored by nursing staff and limited in their actions. This is especially pertinent to residents with dementia who may cause fires without realizing what they have done. Yet these policies are not always standard in other assisted living facilities, thus increasing the risk for fire in those locations.

ELECTRICAL SYSTEMS[2]

According to the National Institute of Occupational Safety and Health (NIOSH), electrical equipment poses a major risk of fire in long term care facilities. The wide range of equipment includes everything from devices used for resident care to cooking equipment in the facilities' kitchens. In the chapter "Recommended Guidelines for Controlling Safety Hazards in Hospitals," NIOSH notes a series of fire safety issues. Although much of the information for the NIOSH report was obtained from studies conducted in hospitals, NIOSH notes that the observations and recommendations can also be applied to other health care settings, including nursing homes. NIOSH cites the following problems with equipment in care settings that pose the highest fire risk:

- three wire plugs attached to two wire extension cords;
- grounding prongs that are bent or cut off;
- ungrounded appliances resting on metal surfaces;
- extension cords with improper grounding;
- cords molded to plugs that are not properly wired;
- ungrounded and multiple plug spiders found in office areas and nursing stations; and
- personal electrical appliances such as radios, coffeepots, fans, power tools, and electric heaters brought by workers from home.

COMPRESSED GASES AND FLAMMABLE LIQUIDS[2]

Nursing homes and some assisted living facilities use various gases for anesthetics, medications, and cleaning. Because some compressed gases are flammable and all are under pressure, they must be handled with extreme care. According to NIOSH, gas cylinders and the gases they contain pose serious fire hazards.

Compressed gases used frequently in long term care settings include acetylene, ammonia, anesthetic gases, hydrogen, oxygen, and nitrous oxide (which can aid combustion and fuel a fire). Acetylene, and hydrogen are flammable, as are the anesthetic gases. Although oxygen and nitrous oxide are

---

[2]This section is drawn from information in Reference 50.

labeled as nonflammable, they are oxidizing gases that will aid combustion. The use of oxygen in resi dents' areas is an obvious fire hazard. Fires can occur in an oxygen enriched atmosphere because of resident smoking, electrical malfunctions, and the use of flammable liquids. Typically, signs of "O_2 in Use" are posted on the door to the room.

Gases stored in cylinders near steam pipes, water pipes, boilers, flammable solvents, combustible wastes, electrical connections, open flames, or other sources of heat or potential heat pose the highest fire risk.

Certain liquids pose similar risks in the gaseous state. Vapors from these liquids can form an ignitable mixture with air. These vapors, and often the liquids themselves, can be ignited by just a spark. NIOSH identifies the following liquids (among others), many of which may be prevalent in nursing home settings, as flammable or combustible and potentially dangerous: most alcohols, acetone, ethyl ether, lubricating oils, ethylene glycol, carbolic acid, cleaning solvents, and most oil based paints. Such liq uids pose additional threats if they are not stored or transported properly.

In a fire situation, flammable gases and liquids yield an additional risk of fire injury and fatality, as they can propel and expand fires and make them more difficult to extinguish. Flammable gases from com busting or exploding cylinders can pose a distinct challenge to the fire service. In addition to spreading, fueling, and increasing the size and scope of a fire, some gases and vapors from liquids can asphyxiate long term care facility residents, rendering them unconscious and making evacuation difficult.

ARCHITECTURAL ISSUES AND BUILDING LAYOUT

The risk of fire fatality in long term care facilities is often higher than in residences because of the architecture and layout of the facilities. Many older buildings have walls, long hallways, multiple floors, and elevators that pose distinct risks to the fire service when evacuating these facilities during fires, since most residents require help during evacuations. Recognizing these architectural deficien cies, many recently built long term care facilities are designed with greater ambience for the residents and easier egress during emergencies.

Although many of these architectural concerns also apply to multifamily units (e.g., apartments) in general, the physical issues associated with older adults compound the fire risk. Regulations do not mandate that walls between residents' rooms resist the passage of smoke; thus, residents in rooms adja cent to fires are at risk from smoke inhalation. [Ref. 45] Hallways fill with smoke, reducing visibility and trapping occupants. Long hallways require occupants and firefighters to travel further to reach fresh air. An occupant may encounter fresh air when beginning his escape during an evacuation, but the same hallway can rapidly fill with smoke, trapping and disabling the occupant.

Many long term care facilities have multiple floors. Multiple floors can lengthen the time needed to effectively evacuate occupants. Elevators are not used during fire conditions, so occupants have to be evacuated through stairwells. Stairwell evacuation may be possible for mobile residents, but disabled or bedridden victims need assistance, which can significantly increase evacuation time and rapidly deplete resources.

Many long term care facilities with multiple floors also have elevators. As with other multistory buildings, elevators can malfunction during a fire and trap victims inside, presenting another challenge to responding firefighters. Elevator shafts can act as chimneys, rapidly moving fire, smoke, and other toxic gases vertically from lower floors to upper floors.

SPRINKLERS AND SMOKE ALARMS

Sprinklers are standard in long term care facilities. However, according to the GAO, older nursing homes are often allowed to operate without sprinklers if they are constructed with noncombustible materials. Even when sprinklers are not in place, federal standards do not require smoke alarms in individual nursing home rooms. [Ref. 45]

Chapter 6 - HOME HEALTH AND HOSPICE CARE

For sick and disabled older adults, nursing homes and assisted living facilities are not the only care options. Home health care—managed by home care agencies, attended by visiting nurses, or managed by the residents themselves—is a desirable option for many older adults seeking privacy and familiarity and who can afford it. The services of hospices (agencies that provide end of life care to patients) are primarily offered in resident homes as well.

Although the comfort and convenience of home health care and hospice care are often attractive to older adults, they too pose fire risks to residents and challenges to the fire service in combating fires in such settings.

Characteristics of Home Health and Hospice Care

According to the National Association of Home Care: [Refs. 51 and 52]

- There are more than 20,000 home health care agencies, 2,265 of which are hospices.
- Residents spent $36.6 billion on home health care in 2000.
- The average cost of a home health care visit in 2000 was $100, and the average yearly expenditures were approximately $5,400.
- Home health care workers range from registered nurses, licensed private nurses, physical therapists, home care aides, occupational therapists, and social workers, to other staff.

Home health care is provided to 5 percent of those 65 to 74 years old, 16 percent of those 75 to 84 years old, and 28 percent of those 85 years and older. [Ref. 43] Most older home health care users are women (64.4 percent) and white (83.8 percent white or other races, 10.8 percent black, and 5.4 percent Hispanic). [Ref. 51]

Of the individuals who receive home health care, the most common illnesses treated are circulatory system disease (33.5 percent), heart disease (18.9 percent), injury and poisoning (14.7 percent), musculoskeletal system and connective tissue disease (13.3 percent), and respiratory system disease (11.1 percent). [Ref. 51] Cancer is the primary disease among residents receiving hospice care. [Ref. 52] As older adults represent the majority of home health care recipients, these statistics offer insight into the health problems of older adults receiving home health care.

Due to the lower cost of home health care compared to hospitals and nursing homes, people with lower incomes are more likely to rely on home health care than other care options. Of the recipients of home health care, 33 percent of residents are in the lowest income brackets. [Ref. 43]

Of all home health care users over age 50, 74 percent required assistance with ADLs and IADLs. The most common ADLs with which home health care residents need assistance include bathing (48 percent), dressing (42 percent), transferring or moving from one place to another (33 percent), and toileting (24 percent). The most common IADLs with which home health care residents need assistance include performing light housework (35 percent), walking (31 percent), preparing meals (21 percent), and taking medications (19 percent). [Ref. 43]

Fire Risk Factors in Home Health and Hospice Care

Similar to long term care facilities, the health of residents in home health care settings is perhaps the most integral fire risk factor. Gases and liquids stored in the home and electrical devices pose similar risks. Home health care, however, has several distinct risk factors that differ from long term care facilities. In home health care settings, factors such as smoking, the level of home fire safety measures in effect, the presence and education of home health staffers, and socioeconomic factors can have an adverse effect on the risk of fire.

RESIDENT HEALTH

Perhaps the highest risk of fire is for older adults in home health care situations with physical and mobility impairments. As noted above, the majority of home health care users require assistance with ADLs and IADLs. This need makes responding to and escaping from fires in their homes a serious fire risk. Thirty three percent of residents have difficulty moving from room to room and another 31 percent have difficulty walking at all. [Ref. 43]

INCOME

Income is not a substantial risk factor in long term care facilities, but it is a factor in home health care. Regulation of long term care facilities, both of the structures and of the care provided, ensures relatively uniform service for all residents regardless of income. In home health care situations, where structures are not regulated beyond traditional residential housing fire codes and where the regulation and training of home health care workers is variable, income is a considerable factor. Upgrades to the home to accommodate home health care needs alone can be expensive for individual households. Upgrades to add or improve fire alert systems, add safety devices to their homes, replace old and dangerous appliances, or clean their chimneys, all of which help mitigate or avoid fire hazards, may also be beyond their means.

GASES AND LIQUIDS

As previously discussed, compressed gases and flammable liquids found in nursing homes, assisted living facilities, and hospitals pose fire risks. When these materials are introduced into a private home, the risks are magnified. Most long term care facilities have regulations, policies, and procedures that dictate the proper handling, storage, and use of such materials. Less attention may be paid to such matters in home settings, thus increasing the risk for fire.

Additionally, the risk for fire fatality and injury due to the presence of gases and liquids in the home may also be higher. Punctured cylinders of oxygen can fuel a fire and make it more deadly and damaging. Such materials also pose challenges to the fire service. Although firefighters can expect that long term care facilities house medical gases and flammable liquids, they often have no way of knowing if such materials are present in residential structures. Because home health care situations are not always obvious, the unexpected presence of these gases can increase the danger to firefighters.

ELECTRICAL DEVICES

Medical equipment that requires electricity is often a component of home health care. Unlike long term care facilities, home health care settings are not usually retrofitted with electrical systems that can handle many pieces of equipment and appliances. Medical equipment can overload a residential electrical system and cause a fire.

SMOKING

In most long term care facilities, smoking is prohibited. This is not the case in home health care settings. Smoking is a legal act and, despite the health and fire risks inherent in smoking, residents can choose to do so. Impaired and disabled residents who choose to smoke (and who in long term care facilities would not be permitted to smoke) place themselves at a higher risk of fire fatality and injury.

HOME SAFETY MEASURES

Long term care facilities are often built with fire resistant materials and are required to meet special building and fire codes. No such protocols exist for private residences. Long term care facilities are also designed with fire safety features such as special alert devices and alarms, air ducts that close in the event of fire or smoke, sprinkler systems and tested fire extinguishers, flame retardant mattresses and drapery, and evacuation features. But private homes are rarely outfitted with such safety features.

Kitchens in long term care facilities are often safely separated from the storage of dangerous health care materials. In home health care situations, kitchens can be mere steps away from the room in which an older resident resides, thus making the spread of kitchen fires more likely in home health care settings.

STAFF PRESENCE AND EDUCATION

Individuals employed in the home health care industry can affect a resident's risk for fire, fire fatality, and fire injury. First, the physical presence of health care staff is a factor. In long term care facilities, staff members are usually on hand at every minute and would be in a position to notice and address fire risks, report fires, and assist residents in the event of an emergency. In most situations, home health specialists visit residents' homes periodically, so there is no 24 hour presence. Without a staff member present, a disabled or impaired resident is at a higher risk of fire fatality and injury than in a long term care facility.

The education of private home health care staff regarding safety also varies greatly. Some staff members may understand fire risks and fire prevention techniques, while others may not.

The risk of fire fatality varies greatly by age, gender, and race. When comparing fire rates, the relative risk of a group, such as older adults, is calculated by comparing its rate to the rate of the overall popula tion. The result is a measure of how likely the group will be affected.

Per Capita Rates, Risk, and Fire Casualties

When determining fire risk, many factors come into play. People in the Southeast, the elderly, the very young, and the poor all are at higher risk of fire death than the rest of the population. Males, African Americans, and American Indians have a considerably higher risk of death from fire than does the pop ulation as a whole. These groups have remained at higher risk despite long term reductions in fires and fire casualties.

To account for differences in population group sizes, per capita rates are used. Per capita rates use a common population size, which then permits comparisons between different groups.[1] The most use ful way to assess fire casualties across groups is to determine the relative risk of dying or being injured. Relative risk compares the per capita rate for a particular group (e.g., females) to the overall per capita rate (i.e., the general population). The resulting ratio is a measure of how likely a group is to be affected.

Table 6 provides details on fire fatalities and relative risk for various population groups, including older adults. According to NCHS mortality data, 1,250 older adults died as a result of fire in 2001, account ing for nearly one third of fatalities.[2] Of every 1 million older adults, 35.4 died in fires in 2001, compared to 14 deaths per million for the overall population (including older adults), resulting in a relative risk of 2.5 for older adults.

---

[1] Per capita rates are determined by the number of deaths or injuries occurring to a specific population group divided by the total population for that group. This ratio is then multiplied by a common population size. For the purposes of this report, per capita rates for fire deaths and injuries are measured per 1 million persons. For example, the per capita fire death rate for the total female population is computed from the total number of female fire deaths (1,552) divided by the total female population (145,241,949) multiplied by 1,000,000 persons. This rate is equivalent to 10.7 deaths per 1 million population.

[2] From USFA sponsored analysis of NCHS 2001 Multiple Cause of Death Public Use data files. Based on this NCHS mortality data, there were 4,007 fire related deaths in 2001, where both age and race were recorded. These include all deaths in which fire or explosion was the underlying cause of death or was a contributing factor in the chain of events leading to death. This latter condition is an expanded approach to capturing fire and fire related deaths. With this current approach, deaths where such exposures were a contributing factor (i.e., the death may not have occurred without the exposure) can be captured. Previous data and methodologies resulted in the ability to capture only those deaths that directly resulted from the exposure to fire and fire products and yielded more conservative numbers. The most conservative definition (fire and flame only, Inter national Classification of Disease codes X00 09) yields 3,326 fire related deaths for 2001. The codes included in this report's mortality statistics are F63.1, W39 W40, X00 X09, X75 76, X96 97, Y25 26, and Y35.1.

The last column in each section in Table 6 provides the relative risk for fire deaths for each category of older adults relative to the general population. The relative risk for the overall population is 1.0 and is used as a baseline for comparisons with distinct subgroups. The relative risk for the overall popula tion encompasses all fire fatalities for all age groups.

Table 6. Fire Fatalities and Relative Risk for Older Adults

Gender/Race	2001 Population	2001 Fire Deaths	Death Rate (per million)	Relative Risk
Total Population				
Total	285,317,559	4,007	14.0	1.0
Male	140,075,610	2,455	17.5	1.2
Female	145,241,949	1,552	10.7	0.8
White	230,664,347	2,908	12.6	0.9
Africa American	36,283,895	1,006	27.7	2.0
American Indian	2,713,047	49	18.1	1.3
Asian/Pacific	11,602,700	44	3.8	0.3
White Male	113,863,214	1,777	15.6	1.1
African American Male	17,256,399	616	35.7	2.5
American Indian Male	1,357,962	33	24.3	1.7
Asian/Pacific Male	5,609,267	29	5.2	0.4
White Female	116,801,133	1,131	9.7	0.7
African American Female	19,027,496	390	20.5	1.5
American Indian Female	1,355,085	16	11.8	0.8
Asian/Pacific Female	5,993,433	15	2.5	0.2
All Older Adults (Age 65+)				
Total	35,353,266	1,250	35.4	2.5
Male	14,619,070	638	43.6	3.1
Female	20,734,196	612	29.5	2.1
White	31,182,858	942	30.2	2.2
African American	2,915,629	283	97.1	6.9
American Indian	154,885	8	51.7	3.7
Asian/Pacific	904,344	17	18.8	1.3
White Male	12,972,922	473	36.5	2.6
African American Male	1,112,495	151	135.7	9.7
American indian Male	66,748	5	74.9	5.3
Asian/Pacific Male	384,994	9	23.4	1.7
White Female	18,209,936	469	25.8	1.8
African American Female	1,803,134	132	73.2	5.2
American Indian Female	88,137	3	34.0	2.4
Asian/Pacific Female	519,350	8	15.4	1.1

Table 6. Fire Fatalities and Relative Risk for Older Adults (cont'd)

Gender/Race	2001 Population	2001 Fire Deaths	Death Rate (per million)	Relative Risk
Age 65–74				
Total	18,322,480	471	25.7	1.8
Male	8,301,935	271	32.6	2.3
Female	10,020,545	200	20.0	1.4
White	15,908,293	342	21.5	1.5
African American	1,655,175	119	71.9	5.1
American Indian	95,278	5	52.5	3.7
Asian/Pacific	549,906	5	9.1	0.6
White Male	7,283,120	190	26.1	1.9
African American Male	685,757	75	109.4	7.8
American Indian Male	43,615	3	68.8	4.9
Asian/Pacific Male	238,155	3	12.6	0.9
White Female	8,625,173	152	17.6	1.3
African American Female	969,418	44	45.4	3.2
American Indian Female	51,663	2	38.7	2.8
Asian/Pacific Female	311,751	2	6.4	0.5
Age 75–84				
Total	12,582,484	493	39.2	2.8
Male	4,996,556	243	48.6	3.5
Female	7,585,928	250	33.0	2.3
White	11,261,249	381	33.8	2.4
African American	934,936	100	107.0	7.6
American Indian	45,354	2	44.1	3.1
Asian/Pacific	279,498	10	35.8	2.5
White Male	4,498,839	188	41.8	3.0
African American Male	337,353	49	145.2	10.3
American Indian Male	18,504	1	54.0	3.8
Asian/Pacific Male	117,658	5	42.5	3.0
White Female	6,762,410	193	28.5	2.0
African American Female	597,583	51	85.3	6.1
American Indian Female	26,850	1	37.2	2.7
Asian/Pacific Female	161,840	5	30.9	2.2

Table 6. Fire Fatalities and Relative Risk for Older Adults (cont'd)

Gender/Race	2001 Population	2001 Fire Deaths	Death Rate (per million)	Relative Risk
Age 85+				
Total	4,448,302	286	64.3	4.6
Male	1,320,579	124	93.9	6.7
Female	3,127,723	162	51.8	3.7
White	4,013,316	219	54.6	3.9
African American	325,518	64	196.6	14.0
American Indian	14,253	1	70.2	5.0
Asian/Pacific	74,940	2	26.7	1.9
White Male	1,190,963	95	79.8	5.7
African American Male	89,385	27	302.1	21.5
American Indian Male	4,629	1	216.0	15.4
Asian/Pacific Male	29,181	1	34.3	2.4
White Female	2,822,353	124	43.9	3.1
African American Female	236,133	37	156.7	11.2
American Indian Female	9,624			
Asian/Pacific Female	45,759	1	21.9	1.6

Note: Relative risk numbers have been rounded.

Sources: References 53 and 59.

Risk of Fire Fatality by Demographic Segment

Generally speaking, as Americans age, their risk of dying in a fire increases. Similar to the relative risk for the overall population, men and African Americans are at a higher risk of death—especially among the "oldest old"—than women and whites. This information is especially important in targeting public education programs to appropriate audiences and providing the fire service with the information needed to combat the fire problem of older adults.

AGE

The relative risk of fire fatality grows as older adults continue to age. The "youngest old," people aged 65 to 74 years, were 1.8 times more likely than the overall population to die in a fire in 2001, according to the data in Table 6. Adults aged 75 to 84 were 2.8 times more likely to die in fires, while the "oldest old," those adults 85 years and over, were 4.6 times more likely to die. When age is compared with such factors as gender, however, the rate of increase in relative risk becomes even more dramatic.

The rate of increase for relative risk and fatality among the age segments of the older population is similar to increases in prescription drug usage, poverty, and disabilities. Although there is not enough research to make a correlation between the increase in these risk factors by age and the increase of

relative fire risk by age, it can be inferred that all of these factors together, along with the general deteri
oration of health and mobility of adults as they age, affect both relative risk and fatality rates.

GENDER

The risk of fire is not uniform across genders. For the population as a whole, men are more likely than women to be victims of fires. This disparity not only holds for older adults, but increases with age. Males 65 years and older were 3.1 times more likely to die in a fire than the overall population, compared to a relative risk of 2.1 for older women (Table 6). While the relative risk of fire death increases steadily to 3.7 for older women as they age, it escalates most rapidly for older men to 6.7.

The data raise other questions about senior Americans. Women have a longer life expectancy than men, but they also have higher rates of disability and poverty, which on the face of it would indicate they are at a higher risk for fire death. Further research on why men have a higher relative risk of fatality in fire is warranted.

RACE

Older African Americans are at a higher relative risk for fire fatality than all other races, with an overall relative risk of fatality 6.9 times higher than the overall population. Meanwhile, the data in Table 6 show that the risk for older whites is much lower: they are only 2.2 times more likely to die in fires than the overall population. In comparison, older American Indians have a relative risk 3.7 times higher than the overall population, while older Asians/Pacific Islanders have a relative risk of just 1.3.

Although the relative risk of death from fire for white older adults increases in a pattern similar to the increase in risk of the overall older population, for African Americans it increases at a much higher rate. And again, the oldest African American men fared much worse than all others. For example, among Americans 85 years and over, African American males are 21.5 times more likely than the over all population to die in a fire, while white males are 5.7 times more likely.

Older African Americans are often in poorer health than older whites, and they have a higher rate of poverty than whites. These are possible reasons for the differences between the older white adult's relative fire fatality risk and that of the older African American.

NFIRS is a USFA fire incident database. NFIRS tracks residential fire fatalities and injuries, the nature and causes of such fatalities and injuries, and other factors regarding fires. NFIRS offers a relative repre sentation of fires and their causes, as not all fires are reported to NFIRS. Of the 537 residential structure fire fatalities reported to NFIRS for adults 18 and over in 2002,[1] 182 of them (or 34 percent) were adults 65 years and older. Of the reported 3,101 residential structure adult fire injuries, 428 (or 14 percent) were among adults 65 years and older (Table 7).

Table 7. 2002 Adult Fire Casualties by Age Group
(percent)

Age Group	Fatalities	Injuries
18 64	66	86
65+	34	14
65+ Subgroups		
65 74	36	42
75 84	40	43
85+	24	15

The analysis of the fire fatality and injury characteristics are based on valid incident data from NFIRS. Incident data are considered valid when data fields contain validated codes in the incident reporting data. Blank and unknown data are not considered valid.

Fatalities and Injuries by Gender

More elderly men died in residential fires than did elderly women, but more women were injured (Table 8). While this may seem like a contradiction, the research noted earlier suggests that more men are exposed to residential fires overall because a higher proportion of elderly women enter nursing homes and assisted living facilities (data for such facilities are not included in this residential analysis), while men primarily stay in residential communities. Older women are more likely to have disabilities and impairments, so the ones who do live in residential structures may be more likely to be injured. In the "oldest old" age group of 85 years and over, there are far more women alive than men.

[1]The information in this chapter is based on NFIRS 5.0 data for 2002 in residential structures. All fatality calculations in the tables and figures herein are based on 537 valid responses; the injury calculations are based on 3,101 valid responses.

Table 8. Casualties by Gender and Age Group
(percent)

Age Group	Fatalities		Injuries	
	Male	Female	Male	Female
18 64	61	39	56	44
65+	52	48	38	62
65+ Subgroups				
65 74	58	42	50	50
75 84	47	53	30	70
85+	50	50	30	70

Fatalities and Injuries by Time of Day

An analysis of NFIRS data indicates that residential fatalities and injuries among older adults varied by time of day and that characteristics of elderly fatalities and injuries by time of day differed from those of the 18 to 64 year old population.

Older adults experienced fewer fatalities in the evening and early morning hours than the 18 to 64 year old population, but experienced more fatalities in the midmorning and early afternoon hours (Figure 7). Similarly, older adults suffered fewer fire related injuries during the late night and early morning hours than the 18 to 64 year old population, and more during the midmorning to early afternoon (Figure 8). It can be inferred that the reason for this difference is that older adults are more likely to be at home during the traditional workday than the 18 to 64 year old population. It also may be a reflection that the mundane daily living activities, such as cooking, are more hazardous for older adults.

As shown in Figure 9, fatalities for older adults were at their highest point during the early morning hours, and injuries peaked in the early evening. Early morning fire fatalities were predominately

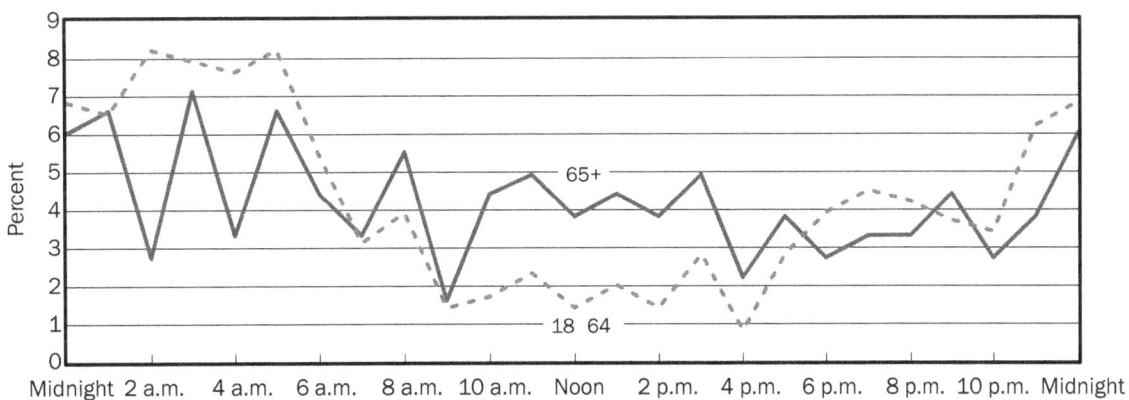

Figure 7. Adult Fire Fatalities by Time of Day

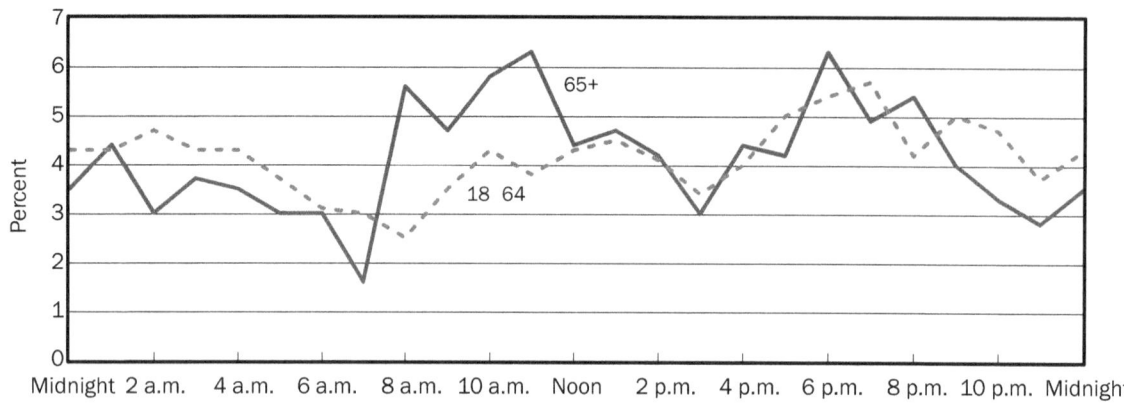

Figure 8. Adult Fire Injuries by Time of Day

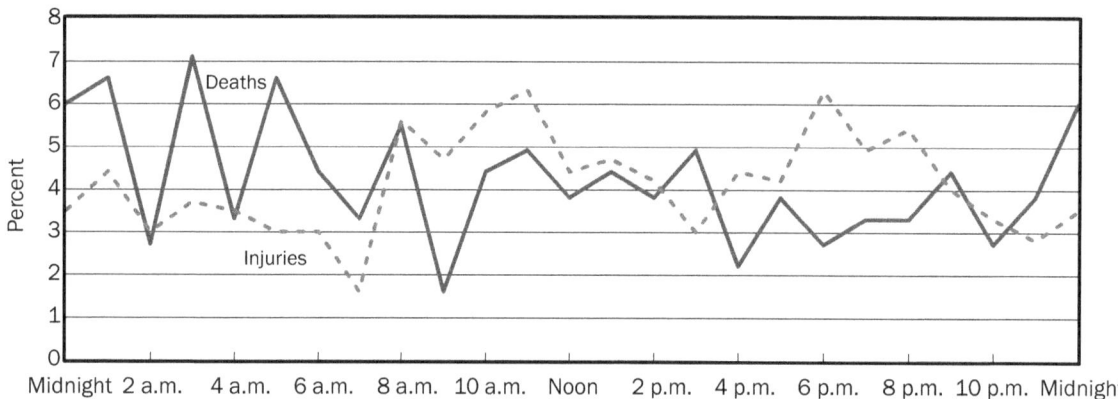

Figure 9. Older Adult Fire Fatalities Versus Injuries by Time of Day

caused by smoking and heating fires, with incendiary and suspicious (arson) and open flame (e.g., candles or matches) fires as contributors. Sleeping and impaired older adults may not realize their homes are on fire before it is too late to escape. The high percentage of injuries in the early evening (and low percentage of fatalities during that time) is due to cooking caused fires. It is likely that older adults involved in such fires are better able to evacuate their homes than sleeping, impaired adults, and they may be aided by home health care workers or family members during the early evening hours.

Fatalities and Injuries by Month

By month, fatalities and injuries for older adults generally followed the same trends as the 18 to 64 year old population. As Figures 10 and 11 indicate, injuries and fatalities from residential fires were higher in the colder winter and fall months than in summer months, with the highest number of fatalities and injuries for older adults occurring in December and March, respectively. This increase in residential structure fire fatalities and injuries is due to the increase in home heating fires during those months.

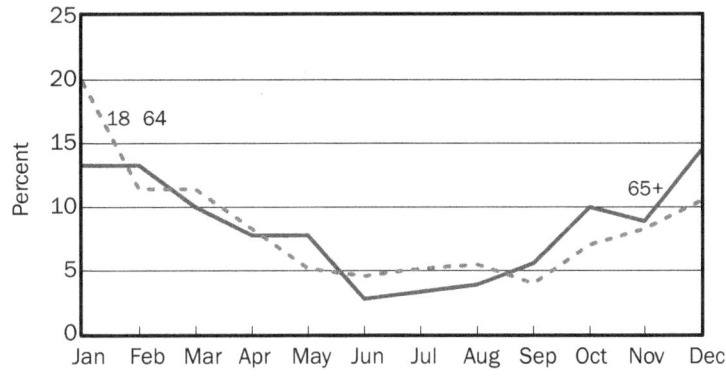

Figure 10. Fire Fatalities by Month

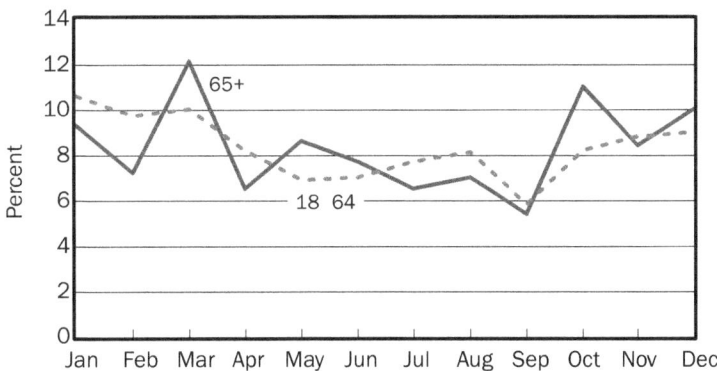

Figure 11. Fire Injuries by Month

Fatalities and Injuries by Location

Of the older adults who died or were injured in a residential fire, most were located in bedrooms, kitch ens, and common areas at the time of fatality or injury. About half of the fatalities to older adults (48 percent) were in bedrooms. The greatest number of older adult casualties were in bedrooms designed for fewer than five occupants. These statistics (Table 9) follow a similar pattern for the locations of adults aged 18 to 64 when they were killed or injured in a fire (Table 10).

Table 9. Leading Locations at Time of Casualty
for Adults Aged 65 and Older

Location	Number	Percent
Fatality		
Bedroom for <5 persons	17	37.0
Other Functional Area	5	10.9
Bedroom for >5 persons	5	10.9
Kitchen	5	10.9
Injury		
Bedroom for <5 persons	35	30.7
Common Area/ Living Room	14	12.3
Kitchen	12	10.5
Egress, Exit	10	8.8
Other Functional Area	10	8.8

Table 10. Leading Locations at Time of Casualty
for Adults 18 to 64

Location	Number	Percent
Fatality		
Bedroom for <5 persons	48	53.3
Bathroom	9	10.0
Kitchen	9	10.0
Common Area/Living Room	7	7.8
Injury		
Bedroom for <5 persons	161	29.3
Kitchen	61	11.1
Common Area/ Living Room	49	8.9
Egress, Exit	31	5.6

Activity When Killed or Injured

Table 11 indicates that adults were most likely to have been sleeping at the time of the fire fatality. The percentages for adults aged 18 to 64 and those 65 and over who were sleeping are nearly identical (40 and 39 percent, respectively). About the same percentage of older adults and adults aged 18 to 64 were killed while escaping a residential structure fire (33 percent). With increasing age, however, the data indicate that a smaller percentage of older adults are sleeping at the time of their deaths; the very oldest fatalities were more likely to be trying to escape.

Table 11. Activity of Adult When Killed or Injured by Fire (percent)

Activity When Killed	18–64	Older Adults			
		All Older Adults	65–74	75–84	85+
Fatalities					
Other activity	3.5	6.5	0.0	5.9	16.7
Escaping	33.3	32.5	32.0	29.4	38.9
Rescue attempt	2.8	1.3	4.0	0.0	0.0
Fire control	2.1	7.8	8.0	8.8	5.6
Returning to vicinity of fire before control	3.5	1.3	0.0	2.9	0.0
Returning to vicinity of fire after control	0.0	0.0	0.0	0.0	0.0
Sleeping	40.3	39.0	48.0	38.2	27.8
Unable to act	11.1	6.5	0.0	11.8	5.6
Irrational act	3.5	5.2	8.0	2.9	5.6
Injuries					
Other activity	7.5	8.2	9.5	5.8	10.5
Escaping	20.4	32.3	34.5	32.0	26.3
Rescue attempt	7.6	3.1	2.6	2.9	5.3
Fire control	41.1	27.6	25.9	30.1	26.3
Returning to vicinity of fire before control	5.6	6.6	9.5	4.9	2.6
Returning to vicinity of fire after control	0.5	1.2	0.9	1.0	2.6
Sleeping	11.5	11.7	11.2	13.6	7.9
Unable to act	1.6	7.0	3.4	8.7	13.2
Irrational act	4.3	2.3	2.6	1.0	5.3

Note: Percentages may not total to 100 percent due to rounding.

For injuries, a much smaller percentage of older adults were injured by trying to control the fires (28 percent) than the 18 to 64 year old population (41 percent), but a higher percentage of older adults were injured while attempting to escape fires (32 percent versus 20 percent).

Causes of Fire in Fatality and Injury Incidents

Smoking, open flame, heating, and suspicious acts, in that order, caused more residential structure fires that resulted in fatalities among older adults than any other fire cause. Cooking, open flame, smoking, and heating caused more residential structure fires that resulted in injuries than any other fire cause.

As Table 12 shows, the causes of fires that resulted in fatalities and injuries among older adults were similar to the causes of fires that killed and injured adults aged 18 to 64 years old. Older adults, however, were slightly more likely to die or be injured in smoking related fires than younger adults, and these younger people were more likely to die or be injured in fires that were caused by suspicious circumstances.

Table 12. Causes of Adult Fire Casualties (percent)

Cause	18–64	Older Adults			
		All Older Adults	65–74	75–84	85+
Fatalities					
Incendiary, Suspicious	24.4	10.5	13.8	8.7	10.0
Children Playing	1.8	0.0	0.0	0.0	0.0
Smoking	22.6	24.2	31.0	26.1	10.0
Heating	12.5	15.8	13.8	10.9	30.0
Cooking	3.0	5.3	3.4	4.3	10.0
Electrical Distribution	3.0	5.3	3.4	6.5	5.0
Appliances	3.0	3.2	0.0	6.5	0.0
Open Flame, Ember, Torch	9.5	17.9	24.1	15.2	15.0
Other Heat, Flame, Spark	15.5	12.6	6.9	17.4	10.0
Other Equipment	3.0	2.1	0.0	0.0	10.0
Natural	0.6	2.1	3.4	2.2	0.0
Exposure	1.2	1.1	0.0	2.2	0.0
Injuries					
Incendiary, Suspicious	13.0	5.3	5.5	4.2	7.5
Children Playing	2.4	0.0	0.0	0.0	0.0
Smoking	10.2	14.3	14.7	16.7	7.5
Heating	7.2	11.4	14.7	8.3	10.0
Cooking	26.9	33.9	32.1	36.5	32.5
Electrical Distribution	2.7	3.3	2.8	2.1	7.5
Appliances	6.2	5.3	4.6	7.3	2.5
Open Flame, Ember, Torch	19.6	15.5	14.7	13.5	22.5
Other Heat, Flame, Spark	8.9	8.2	6.4	9.4	10.0
Other Equipment	1.5	1.6	2.8	1.0	0.0
Natural	0.9	1.2	1.8	1.0	0.0
Exposure	0.5	0.0	0.0	0.0	0.0

Note: Percentages may not total to 100 percent due to rounding.

Older adults were more likely to die in fires resulting from open flames than were adults aged 18 to 64, and more likely to be injured in cooking fires than adults aged 18 to 64.

Fire education and fire prevention programs for older adults can help America's seniors address fire risks and take necessary precautions to reduce their risk of fire injury and fatality. In addition, these programs can help the fire service more easily save the lives of older adults who know and follow fire evacuation and safety procedures.

Although fire prevention programs for children and young people abound, many fire departments—and even organizations that focus solely on elder needs—do not offer education and prevention pro grams for the older adult. Several programs, however, are exemplary in their approaches and breadth. Five such programs are summarized in this chapter.

Fire Safety Campaign for People 50-Plus

In the summer of 2004, USFA launched the FIRE SAFETY CAMPAIGN FOR PEOPLE 50-PLUS, a fire education campaign guide focused on provid ing USFA partners with the tools and resources to implement public education campaigns for older Americans in local communities. [Ref. 54]

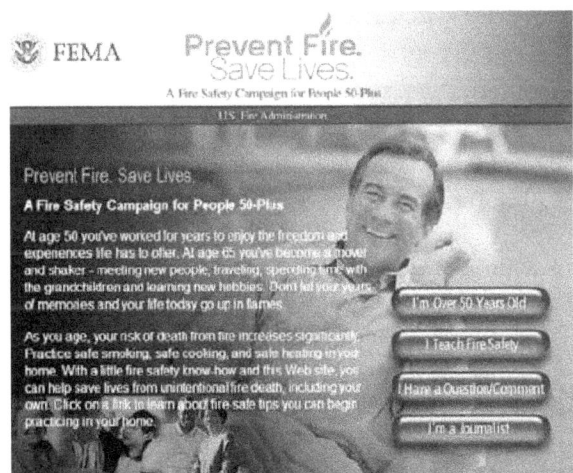

The campaign template encourages people age 50 plus to practice safe smoking, cooking, and heating behaviors, and emphasizes the importance of maintaining smoke alarms, designing and practicing a fire escape plan, and installing home fire sprinklers.

The campaign guide—geared toward fire departments and other organizations with prevention goals—provides a "how to" introduction on implementing a program, promotional materials such as fact sheets and fliers, and sample media materials, such as a news release and radio script. The guide also provides suggestions for working within a community and working with the media. Spanish language campaign materials are also suggested.

USFA has provided its partners with a CD ROM containing campaign materials, and materials are also available for download online at www.usfa.fema.gov/50Plus.

Florida Injury Protection Program for Seniors

The FLORIDA INJURY PROTECTION PROGRAM FOR SENIORS (FLIPS), coordinated by the Florida State Department of Elder Affairs, has been studied and emulated by other States, fire departments, and even governing bodies outside of the United States, according to the program's coordinator. [Ref. 55]

FLIPS has forged partnerships with the Florida Department of Health, Department of Children and Families, Florida State Fire College, Office of the State Fire Marshal, and other State agencies to plan, develop, evaluate, and maintain statewide injury prevention programs for seniors.

 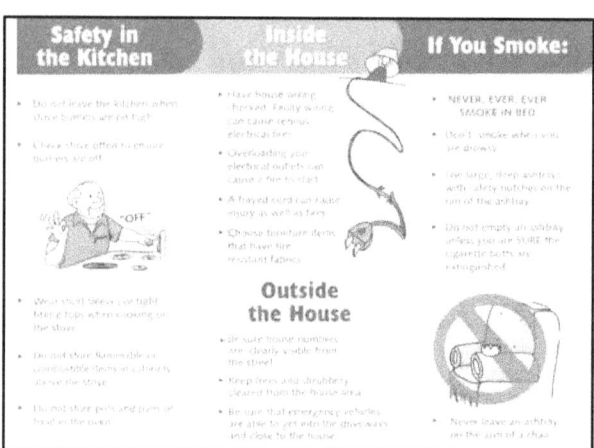

Fire prevention is an integral component of FLIPS. In addition to publishing a widely distributed bro chure, FLIPS is reaching seniors through a statewide speakers bureau where industry leaders and offi cials speak to older adults about risks, periodically providing them with free smoke alarms, and offer ing information about necessary home improvements.

In addition to reaching seniors directly, FLIPS organizes half day workshops on "Fire Safety for Seniors" for home health care nurses, home aides, social workers, hospice personnel, fire inspectors, adult abuse workers, private home health care agency personnel and hospital personnel.

Another component to FLIPS is the prevention of falls, a key factor in mobility in older adults that, as previously discussed, is a major fire risk factor for older adults. This information can be found at http://www.doh.state.fl.us/workforce/ems1/InjuryPrevention/flips.html.

Fire Safety in Long-Term Care Facilities

A joint program of the American Health Care Association (AHCA) and the National Center for Assisted Living (NCAL), FIRE SAFETY IN LONG-TERM CARE FACILITIES is a 4 hour comprehensive course geared toward educating personnel in long term care facilities about the fire risks to older adults and how to safely address fire situations. [Ref. 56] The goals of FIRE SAFETY IN LONG-TERM CARE FACILITIES are to

- identify the need for special fire safety evaluations, fire prevention, and emergency planning;
- identify the basic components and behavior of fire;
- identify the characteristics of residents and staff that affect fire safety and emergency planning;
- identify proper procedures for fire emergency planning and fire drills;
- identify specific fire hazards and procedures for correcting them; and
- identify the function, proper use, and maintenance of fire safety devices.

The FIRE SAFETY program consists of an instructor manual that details how to plan and run work shops on fire safety, two videos, and participant course books that include exercises, checklists, and other pertinent information for personnel. The program consists of six topical modules: Need for Fire Safety, Fire Behavior, Human Fac tors in Fire Safety, Fire Emergency Planning, Fire Hazards, and Fire Safety Devices. Participants are tested to verify their comprehension of the detailed material and case studies presented. The American College of Health Care Administrators has information available on its Web site at http://portal.achca.org/PortalTools/Shopper/ProductDetail.cfm?ProdCompany Passed=ACH&ProdCdPassed=ACH%2DAH%2DFSI.

Remembering When

Organized by the NFPA and the CDC, REMEMBERING WHEN is a comprehensive fire and fall education and prevention program adopted by many fire departments throughout the United States. [Ref. 57] REMEMBERING WHEN provides detailed instructions to fire departments and other entities about the best methods to reach the senior population, including publicity, meetings, and presentations. The key messages of REMEMBERING WHEN are

- provide smokers with large, deep ashtrays;
- give space heaters space;
- be kitchen wise;
- stop, drop, and roll;
- smoke alarms save lives;
- plan and practice your escape from fire;
- know your local emergency number; and
- plan your escape around your abilities.

REMEMBERING WHEN gives participating agencies a detailed program book outlining effective group presentations and techniques for home visits and the installation of smoke alarms. Additionally, departments receive support materials such as message cards, trivia game cards, nostalgia cards, home safety checklists, fact sheets, and brochures.

A presentation developed by the NFPA and CDC can found on the NFPA Web site at http://www.nfpa.org/itemDetail.asp?categoryID=203&itemID=17832&URL=Lear ning/Pub lic%20Education/Remembering%20When.

Senior Safety

The American Burn Association's free SENIOR SAFETY program offers a comprehensive, step by step guide for communities interested in fire and burn prevention programs for older adults. [Ref. 58] Recog

nizing that conveying fire risks and prevention strategies to seniors can be more difficult than educating the general population, the program attempts to

- assess factors among elders that affect their ability to learn;
- diagnose special barriers to learning;
- plan modification in learning conditions;
- implement new learning; and
- evaluate educational outcomes of learning.

Released in February 2003 during National Burn Association Week, the program provides information on fire and burn risks, different types of burn injuries, fire and burn prevention in home health care settings, and information for older adults living independently. A helpful Fire/Burn Home Inspection Instrument checklist provides seniors with a simple guide to reducing fire and burn risks in their homes.

The program provides suggestions for communities about working with the media to publicize the programs to the older adult market and furnishes posters to educate older adults about the fire risks inherent in smoking. SENIOR SAFETY recommends the following crucial elements:

- morning sessions, when energy levels are expected to be highest;
- the establishment of a warm, accepting atmosphere;
- sessions that are conducted in 1 hour or less; and
- a classroom setting that is convenient, easily accessible, and ideally situated in an area that is familiar to participants.

The complete SENIOR SAFETY campaign kit is provided at http://www.ameriburn.org/Pre ven/2003Prevention/2003BurnAwarenessKit.pdf.

REFERENCES

Web sites valid as of November 15, 2005

1. U.S. Census Bureau, Population Division, Population Projections Branch, "U.S. Interim Projections by Age, Sex, Race, and Hispanic Origin," Table 2a, 2004.
    http://www.census.gov/ipc/www/usinterimproj

2. Hetzel, Lisa and Anneta Smith, U.S. Census Bureau, "The 65 Years and Older Population: 2000," *Census 2000 Brief*, October 2001.
    http://www.census.gov/prod/2001pubs/c2kbr01-10.pdf

3. Hobbs, Frank B. and Bonnie L. Damon, Bureau of the Census, Economics and Statistics Administration, *65+ in the United States*, April 1996.
    http://www.census.gov/prod/1/pop/p23-190/p23-190.pdf

4. Hobbs, Frank and Nicole Stoops, U.S. Census Bureau, Census 2000 Special Reports, Series CENSR-4, *Demographic Trends in the 20th Century*, November 2002.
    http://www.census.gov/prod/2002pubs/censr-4.pdf

5. *American Baby Boomers*, Mature Market Institute, Metropolitan Life Insurance Company, 2004, based on U.S. Census data.

6. Kinsella, Kevin and Victoria A. Velkoff, U.S. Census Bureau, Series P95/01-1, *An Aging World: 2001 International Population Reports*, 2001.
    http://www.census.gov/prod/2001pubs/p95-01-1.pdf

7. Federal Interagency Forum on Aging-Related Statistics, *Older Americans 2004: Key Indicators of Well-Being*. November 2004. http://www.agingstats.gov/chartbook2004/OA_2004.pdf

8. U.S. Department of Health and Human Services Administration on Aging, *A Profile of Older Americans: 2003*. http://www.aoa.gov/prof/Statistics/profile/2003/2003profile.pdf

9. National Center for Health Statistics, "United States Life Tables, 2002" *National Vital Statistics Report*, Vol. 53, No. 6, Table 12, November 2004.
    http://www.cdc.gov/nchs/data/nvsr/nvsr53/nvsr53_06.pdf

10. National Center for Health Statistics, Data Warehouse, *Leading Causes of Death, 1900–1998*.
    http://www.cdc.gov/nchs/data/dvs/lead1900_98.pdf

11. U.S. Census Bureau, Population Division, Table NA-EST2002-ASRO-01 - Annual Resident Population Estimates of the United States by Age and Sex: April 1, 2000 to July 1, 2002, Release Date: June 18, 2003.
    http://www.census.gov/popest/archives/2000s/vintage_2002/NA-EST2002-ASRO-01.html

12. "Scents Will Not Rouse Us From Slumber, Says New Brown University Study," *Science Daily*, May 18, 2004. http://www.sciencedaily.com/releases/2004/05/040518075747.htm

13. "Sense of Smell Tied to Seniors' Mental Functioning." *Psychology and Aging*, 17:392-404, 2002.

14. Bromley, Steven M., "Smell and Taste Disorders: A Primary Care Approach," *American Family Physician*, January 15, 2000. http://www.aafp.org/afp/20000115/427.html

15. Murphy, C., Schubert, C.R., Cruickshanks, K.J., Klein, B.E.K., Klein, R., and Nondahl, D.M., "Prevalence of Olfactory Impairment in Older Adults," *Journal of the American Medical Association*, 288:2307-2312, 2002.

16. Ahmadi, Reza and Kay Hodson Carlton, WellCome Home Project, Ball State University. http://www.bsu.edu/wellcomehome/safety.html

17. U.S. National Library of Medicine and National Institutes of Health, MedlinePlus Medical Encyclopedia, "Aging Changes in Skin." http://www.nlm.nih.gov/medlineplus/ency/article/004014.htm

18. "Normal Changes in the Aging Eye," VisionConnection. http://www.visionconnection.org/Content/YourVision/TheAgingEye/NormalChangesintheAgingEye.htm

19. "The Four Most Common Causes of Age-Related Vision Loss." VisionConnection, http://www.visionconnection.org/Content/YourVision/TheAgingEye/TheFourMostCommonCausesofAgeRelatedVisionLoss.htm

20. Yablonski, Martin, S., "Loss of Vision in Later Life: A Different Perspective," *Aging and Vision*, Volume 15, Number 1, Spring 2003.

21. Desai, Mayur, et al., *Trends in Vision and Hearing Among Older Americans*, Centers for Disease Control and Prevention, National Center for Health Statistics, Aging Trends No. 2, March 2001. http://www.cdc.gov/nchs/data/agingtrends/02vision.pdf

22. Reese, Shelly, "Research is Needed To Determine Why Children Are Sleeping Through Smoke Alarms," *NFPA Journal*, July/August 2003. http://www.nfpa.org/categoryList.asp?categoryID=724&URL=Publications/NFPA%20Journal®/July%20/%20August%202003/Features

23. *Statistics About Alzheimer's Disease*, Alzheimer's Association, July 2002. http://www.alz.org/AboutAD/statistics.asp

24. "Safety in the Home," *After the Diagnosis*, Alzheimer's Society, January 2002. http://www.alzheimers.org.uk/After_diagnosis/PDF/asafety_home.pdf

25. Hendrickson, Gail, "Depression in the Elderly," *Diseases and Conditions*, discoveryhealth.com, September 2001. http://health.discovery.com/encyclopedias/2844.html

26. "Seniors' Falling Injuries Are Preventable," *WebMD Quick Facts*. http://my.webmd.com/content/article/11/1738_50053.htm

27. Parra, Elizabeth K., and Judy A. Stevens, *U.S. Fall Prevention Programs for Seniors*, Centers for Disease Control and Prevention, National Center for Injury Prevention and Control, Atlanta, Georgia, 2000. http://www.cdc.gov/ncipc/falls/default.htm

28. "Falls Among Older Persons and the Role of the Home: An Analysis of Cost, Incidence, and Potential Savings from Home Modifications," *Public Policy Institute Issue Brief*, AARP, 2001. http://www.aarp.org/research/housing-mobility/accessibility/aresearch-import-788-IB56.html

29. "Alcohol Use and Abuse," *Age Page*, National Institute on Aging and National Institute on Alcohol Abuse and Alcoholism, May 2002. http://www.niapublications.org/agepages/alcohol.asp

30. "Alcohol and Aging," *Alcohol Alert*, No. 40, National Institute on Alcohol Abuse and Alcoholism, April 1998. http://pubs.niaaa.nih.gov/publications/aa40.htm

31. National Center for Chronic Disease Prevention and Health Promotion, Behavioral Risk Factor Surveillance System, Online Prevalence Data, 2002. http://apps.nccd.cdc.gov/brfss/age.asp?cat=AC&yr=2002&qkey=249&state=US

32. U.S. Food and Drug Administration, FDA Consumer Magazine, September–October 1997, revised January 1999, August 2000, November 2002, and September 2003. http://www.fda.gov/fdac/features/1997/697_old.html

33. "Side Effects of 'Everyday' Drugs in Elderly Can Be Mistaken for Dementia," Royal Society of Medicine, September 2000. http://www.rsm.ac.uk/new/pr68.htm

34. Henry J. Kaiser Family Foundation, "Seniors and Prescription Drugs," July 2002. http://www.kff.org/medicare/6049-index.cfm

35. Johnston, C. Bree. "Drugs and the Elderly: Practical Considerations." UCSF Division of Geriatrics Primary Care Lecture Series, May 2001. http://www.ucop.edu/agrp/docs/sf_drugs.ppt

36. Nash, David B., Jennifer B. Koenig, and Mary Lou Chatterton, "Why the Elderly Need Individualized Pharmaceutical Care," Health Focus, National Pharmaceutical Council, April 2000. http://www.npcnow.org/newsroom/factsheets/PDFs/elderlypharmcare_factsheet.pdf

37. Alcohol Alert, No. 27, "Alcohol-Medication Interactions," National Institute on Alcohol Abuse and Alcoholism, January 1995. http://pubs.niaaa.nih.gov/publications/aa27.htm

38. U.S. Department of Labor, Bureau of Labor Statistics, Labor Force Statistics from the Current Population Survey. http://www.bls.gov/cps/cpsaat3.pdf

39. Proctor, Bernadette D. and Joseph Dalaker, U.S. Census Bureau, Current Population Reports, P60-222, Poverty in the United States: 2002, 2003. http://www.census.gov/prod/2003pubs/p60-222.pdf

40. RAND Center for the Study of Aging, Labor and Population Program, Research Brief, The Relationship Between the Socioeconomic Status and Health of the Elderly, 1998. http://www.rand.org/publications/RB/RB5020

41. Centers for Disease Control and Prevention, Early Release of Selected Estimates Based on Data From the 2004 National Health Interview Survey. http://www.cdc.gov/nchs/data/nhis/earlyrelease/earlyrelease200412.pdf

42. U.S. Census Bureau and U.S. Department of Housing and Urban Development Office of Policy Development and Research, American Housing Survey for the United States: 2003 Current Housing Reports, September 2004. http://www.census.gov/prod/2004pubs/H150-03.pdf

43. A Report to the Nation on Independent Living and Disability—Beyond 50.03, AARP Public Policy Institute, 2003. http://research.aarp.org/il/beyond_50_il.html

44. Mathew Greenwald & Associates, "These Four Walls... Americans 45+ Talk About Home and Community," AARP, May 2003. http://www.aarp.org/research/reference/publicopinions/aresearch-import-769.html

45. Nursing Home Fire Safety: Recent Fires Highlight Weaknesses in Federal Standards and Oversight, Report to Congressional Requestors. United States Government Accountability Office, GAO-04-660. July 2004. http://www.gao.gov/new.items/d04660.pdf

46. Jones, A., The National Nursing Home Survey: 1999 Summary, National Center for Health Statistics, Vital Health Statistics, Series 13, No. 152, 2002. http://www.cdc.gov/nchs/data/series/sr_13/sr13_152.pdf

47. Wright, Bernadette, *Assisted Living in the United States*, AARP Public Policy Institute, October 2004.
http://www.aarp.org/research/housing-mobility/assistedliving/
assisted_living_in_the_united_states.html

48. National Center for Assisted Living, *Assisted Living: Independence, Choice and Dignity*, 2001.
http://www.ncal.org/about/alicd.pdf

49. Centers for Medicare and Medicaid Services, *Program Information on Medicare, Medicaid, SCHIP, and Other Programs of the Centers for Medicare and Medicaid Services*, June 2002.
http://new.cms.hhs.gov/TheChartSeries/downloads/sec2_p.pdf

50. Centers for Disease Control and Prevention, National Institute for Occupational Safety and Health, *Guidelines for Protecting the Safety and Health of Health Care Workers*, September 1988.
http://www.cdc.gov/niosh/hcwold3.html

51. National Association of Home Care, *Basic Statistics About Home Care*, November 2001.
http://www.nahc.org/consumer/hcstats.html

52. National Association of Home Care, *Hospice Facts & Statistics*, November 2002.
http://www.nahc.org/consumer/hpcstats.html

53. U.S. Census Bureau, Population Division, Detailed Files for Monthly Population Estimates, 2000 to 2002. http://www.census.gov/popest/archives/2000s/vintage_2002/files/2002RESIDENT2001MONTHS07_12.txt

54. United States Fire Administration, A FIRE SAFETY CAMPAIGN FOR PEOPLE 50-PLUS.
http://www.usfa.fema.gov/50plus/

55. Department of Elder Affairs, Florida Department of Health, FLORIDA INJURY PREVENTION FOR SENIORS. http://www.doh.state.fl.us/workforce/ems1/InjuryPrevention/flips.html

56. American College of Health Care Administrators, FIRE SAFETY IN LONG-TERM CARE FACILITIES.
http://portal.achca.org/PortalTools/Shopper/ProductDetail.cfm?ProdCompanyPassed=ACH&ProdCdPassed=ACH%2DAH%2DFSI

57. National Fire Protection Association, REMEMBERING WHEN.
http://www.nfpa.org/itemDetail.asp?categoryID=203&itemID=17832&URL=Learning/Public%20Education/Remembering%20When

58. American Burn Association, SENIOR SAFETY, February 2003.
http://www.ameriburn.org/Preven/2003Prevention/2003BurnAwarenessKit.pdf

59. U.S. Census Bureau, Population Division, Table NA-EST2002-ASRO-03 - Annual Resident Population Estimates of the United States by Age, Race, and Hispanic or Latino Origin: April 1, 2000 to July 1, 2002, Release Date: June 18, 2003.
http://www.census.gov/popest/archives/2000s/vintage_2002/NA-EST2002-ASRO-03.html